J. Batty Tuke

The Morisonian Lectures

Delivered before the Royal College of Physicians of Edinburgh, Session

1874

J. Batty Tuke

The Morisonian Lectures
Delivered before the Royal College of Physicians of Edinburgh, Session 1874

ISBN/EAN: 9783337269906

Printed in Europe, USA, Canada, Australia, Japan

Cover: Foto ©berggeist007 / pixelio.de

More available books at **www.hansebooks.com**

THE

MORISONIAN LECTURES.

THE

MORISONIAN LECTURES,

DELIVERED BEFORE

THE ROYAL COLLEGE OF PHYSICIANS OF EDINBURGH.

SESSION 1874.

BY

J. BATTY TUKE, F.R.C.P.E., F.R.S.E.

EDINBURGH : OLIVER AND BOYD, TWEEDDALE COURT.

MDCCCLXXV.

REPRINTED FROM THE EDINBURGH MEDICAL JOURNAL, 1874-1875.

MORISONIAN LECTURES.

LECTURE I.

Mr President,—My first duty is to thank the Patron of the Morisonian Lectureship for the honour he has conferred on me in appointing me to deliver prelections on the subject of Insanity. I beg to assure him and the Fellows of this College that, as far as my light goes, I will endeavour to carry out the wishes and intentions of the Founder.

The study of the multiform conditions comprised under the general term Insanity has been approached by the former lecturers in this College from three different positions. That well-beloved physician, William Sellar, was naturally led, by the character of his studies and the inclination of his mind, to regard the subject from the metaphysical standpoint. Dr Arthur Mitchell[1] brought to bear upon the subject his widely-extended means of observation, his extensive literary researches, and his acknowledged powers of generalization, concentration, and criticism. Were it not for our official relationship, I would wish to dilate more fully on the great service which Dr Mitchell's lectures have effected in widening our views on the nature of insanity in all its relations, medical, social, and legal. In David Skae, my much-beloved and respected master, you had a representative psychiatric clinicist. In the posthumous lectures delivered last year by Dr Clouston (a most delicate task, performed in a spirit of extreme conscientiousness), a system of classification was submitted to you, founded mainly on clinical observation. However we may regard the principle on which this system was based, or the weight of argument adduced to substantiate it, we must acknowledge that it was one of the markings in the epoch of alienistic medicine, in which its study was being raised from the indefinite basis of pseudo-metaphysical generalization and landed on the firmer platform of scientific obser-

[1] One of Her Majesty's Commissioners in Lunacy for Scotland.

A

vation. Skae's principles led to the notation of psychical symptoms as indicative of somatic conditions, and in so doing, to accurate observation and the collation of pathogeny, etiology, and pathology. Its author did not advance it as a perfect system, he only regarded it as a schema of observation which would tend towards the establishment of a storehouse of facts, from which future workers might obtain definite data.

On undertaking the duty of preparing this course of lectures, I felt how difficult was the task before me; as I was to address an audience of men practically acquainted with the treatment of insanity, and whose long clinical practice would render it impertinent in me to submit the results of my comparatively limited experience merely for their criticism or comment; for, believe me, Gentlemen, I am fully cognisant of the fact that it is the non-alienistic physician and the general practitioner who see and treat insanity in its most important stages. The only advantage we asylum physicians have, is, that we are forced to direct our attention almost exclusively to particular points and aspects of mental disease, little cognisance of which is taken in systematic books on the practice of physic, compelling us to concentrate our energies in a manner which the exigencies of practice render all but impossible for a member of the general medical public to do.

Keeping this in view, I deliberated long and anxiously whether the present state of our knowledge was such as to warrant an attempt to approach the subject by the avenue of general pathology, and more especially of morbid anatomy. The track appeared so faint, so many apparent difficulties lay in the way, at certain points it became lost in the broad moor of hypothesis; on the other hand, so many converging lanes seemed to lead towards it, such an amount of the traffic of argument, so many toilers at varying intervals were plodding along it, hardening and consolidating the path, as to justify at least an effort to establish pathology and morbid anatomy as the highway to the study of so-called mental disease. Whether the attempt is too ambitious must be left to your verdict;—to my mind its very difficulty warrants the undertaking. In many points of detail I know success is impossible, but on the main principles involved some light may be thrown.

Allow me shortly to point out the preliminary difficulty I have encountered. It is the general non-acceptance of the pathological fact that insanity is an indication of *disease of the brain.* It is only within a comparatively few years that insanity has in theory been admitted to be a symptom of bodily disease. It is true that most writers on the subject have pointed out that psychical disturbances occur along with certain conditions of the general system, but without exception they have magnified the symptom *to the detriment of the efficient cause.* The prominence given to the consideration of the psychical symptom has been such that the symptom has obtruded itself upon the mind of the observer to such a degree, that it has assumed the position of the disease,—a disease, however, which has been ascribed more to perversion of mental function than to somatic changes, and a condition therefore to be studied by the light of abstract psychology. Thus, every book on insanity which has ever been published commences with an analysis of the Psyche, and the invariable result has been that this preliminary study has so worked upon the authors that when they have arrived at the practical part of their work, the physiology and pathology, which had been interwoven with their psychology, have failed to assert themselves in any substantial form, and the classification of so-called mental disease has assumed the shape of arbitrary catalogues of perversions of function framed in accordance with the tone of the writer's mind. I will only instance the brilliant work of Maudsley, in which a strong foundation of physiology is built up, but which foundation is kicked down by the author himself when he treats of the varieties of insanity as a practical physician. He assigns reasons of the most urgent character for regarding insanity as a disease of the brain; yet when he proceeds to nosology, he reverts to psychology, and lays down two great classes, affective or pathetic insanity, and ideational insanity. Further on in this or the next course, I shall have to revert to the principles of classification; I only allude to this particular instance to show how strong the tendency is to regard the symptom in preference to the efficient morbid cause.

There is no stronger evidence of this tendency than can be gained by examination of the British official nosology; for in the work which stands, and has stood since its publication, on the book-shelf of every member of the profession in Great Britain and Ireland—a work compiled by a joint-committee appointed by the Royal College of Physicians of London, and issued by the authority of the Registrar-General—there will be found a list of some nine hundred diseases, a large assortment of poisons, and fifty-seven pages of accidents and malformations under which the body of the Briton is authorized to suffer or die; the mind of the Briton, however, is authorized to suffer from only six "disorders of the intellect;" the idea of somatic disease as associated with insanity being studiously ignored. On what principle the differentiation between a disease and a disorder is founded, or on what pathological principle it is based, it is difficult to say; still, there the opinion stands, expressed by very high authority, that insanity is not a disease of the body, merely a disorder of the intellect.

Far be it from me to underrate the value of the study of psychology in connexion with the study of insanity; the two cannot be disassociated. I only deprecate as a practical physician its study as a primary study, for it is quite subservient to pathology, general and morbid. Skae put it well when he defined insanity to be "a disease of the brain *with mental symptoms.*" This is the light by which we all regard mental disease in practice. The time

has gone by when any apology is needed for treating the subject
from a pathological and consequently anatomical and physiological
point of view; the everyday work of the physician brings with it
evidences of the subjection of the mental faculties to material
conditions, and the whole tenor of modern research is towards the
localization of nervous function, and consequently to definite
localization of morbid conditions.

But as pathologists how are we to regard the brain functions?
Every now and then the coarsely material apothegm of Cabanis has
cropped up on the surface of metaphysical and psychological discussion, that mental manifestations are as much secretions of the brain as
bile is of the liver. I do not propose to dwell long on this question,
for it possesses no anatomical or physiological ground of support.
The cells of the brain are doubtless dependent on blood-supply for
the exercise of their vital properties, in the same way as the cells of
the liver; but here the analogy begins and ends, for in no way can
the results of brain organization and action be claimed as a secretion. It is not that they are immaterial and impalpable, which
renders inapplicable to them the term secretion; it is that the force
generated in the brain controls the results of the organization and
action of the glands, and that without it they would be powerless
and inoperative. We admit that the brain is dependent on
material conditions for the exercise of its function, but on the
exercise of certain of these functions depends the permanent
exercise of the function of every structure of the body. It is the
organ of the body in which is generated a something which in our
ignorance we call a force, which, whether we view it as vitalists or
physicists, is very closely associated with the phenomena of life.
We are about to direct our attention more especially to the physical
causes of the abnormalities of the phenomena of ideation, emotion,
and volition, although we shall find that with them are intimately
associated perversions of the other phenomena of nerve force, sensation, motion, and nutrition. It is utterly opposed to all the ideas
of the *mens medica* that an insane mind can exist in a sane body;
there is no analogy in any other system; and the fact that the
moral and intellectual faculties are not the only ones implicated,
that the disease which evidences itself by perversion of mental
function is invariably attended by changes in the general system,
affords strong and valuable support to the general theory of the corporeal nature of insanity. Doubtless there are many questions apparently involved which do not admit of pathological explanation; I only
propose, however, to treat of those conditions which I believe are
susceptible of it. The extraordinary epidemics of enthusiasm and
emotionalism of the dark ages and of the present day do not come
within the pale of my subject; for, as a physician, I have nothing
to do with the influence the orator exerts on his audience, whether
he speaks from the platform or the pulpit, until the emotions he
excites stimulate brain action beyond the confines of health in the

individual. Until the patient is before us we have no *locus standi*. This may be a narrow view of the position, but it is the one which must be taken by the physician, and until taken up by him, the lunatic must remain on the border-land of medicine, neither quite accepted nor entirely extruded. It is for us to determine whether he is a metaphysical curiosity or a pathological entity.

I am inclined to elaborate Skae's definition, by saying that *insanity consists in morbid conditions of the brain, the results of defective formation or altered nutrition of its substance, induced by local or general morbid processes, and characterized especially by non-development, obliteration, impairment, or perversion of one or more of its psychical functions.*

This definition is so broad as to include many forms of insanity which do not properly belong to my province; I will therefore fence my lectures by stating that the arbitrary line which exists between the temporary insanities characterized by delirium, coma, or convulsions, and the more permanent forms characterized by mania, delusion or dementia, will be, as far as possible, studiously adhered to, *without prejudice, however, to the opinion that they in fact form one inseparable whole.* As the lecturer on surgery must at times touch slightly on the field of the teacher of medicine, and *vice versa*, so must I, if a broad view of the subject is to be taken, occasionally overstep the arbitrary boundary which exists between general and psychiatric medicine.

Adopting this definition as the proposition to be proved, we will proceed to its demonstration by what appears to me a logical sequence. We will consider,—

1st. The Normal Histology of the Brain.

2d. The more recent Advances in Cerebral Physiology.

3d. The Influences inducing and producing Brain Diseases, with the Associated Symptoms.

4th. The Nature of the Lesions found in the Brains of the Insane.

I. *The Normal Histology of the Brain.*

The study of the pathology of the brain was for long retarded by the imperfection of our knowledge of the finer points of its anatomy. It is within comparatively recent times that methods of investigation have been devised by which the intimate relationships of its delicate structures could be viewed under the microscope. This difficulty was overcome by the discovery that its tissues could be submitted to the hardening influence of chemical agents—such as chromic acid, chloride of gold, iodized alcohol, and others—which, whilst rendering them amenable to the section knife, did not produce any material change in their constituent elements. Immediately on this method of research becoming known, a crowd of workers took up the subject of the microscopic anatomy of the brain; on the Continent, this new field attracted numerous investigators, whilst in this country Lockhart Clarke has rendered his name historical from the bright light his labours have thrown on the histological anatomy of the brain and spinal cord.

The statement that no material change in the cerebral tissues results from their subjection to hardening agency, must, of course, be made with reserve. It is not contended that they present the identical appearance of the living structure; that is impossible in any dead tissue, and more especially in those, the soft nature of which renders it impracticable to cut thin sections in the fresh state; it is only contended that the modification is so slight as to render the demonstration reliable, in so far as that the relation of constituents is not materially interfered with, and that no artificial structures are produced.

As succinctly as possible I will endeavour to lay before you the more recently elucidated points of normal cerebral anatomy. We will consider, first, the several constituents separately, and then their relations one to the other. This step is absolutely necessary in order to give weight to any demonstrations of morbid appearances which it may be in my power to submit to you; and further, it is advisable to collect as far as possible the various facts scattered throughout separate monographs in various languages, or buried in the not very accessible Transactions of the Royal Society of London. Those who are specially acquainted with the subject must bear with me, for I am sure it is no disparagement to the anatomical lore of the mass of my hearers to presuppose that they are not closely conversant with the details of that department of the science which appears to have but slight bearing on general medicine. Nor would I enter on these details, were it not that I hope to give you ocular demonstration of morbid appearances in each and all of the cerebral structures in insane brains.

What I desire most particularly to direct your attention to, is the histological anatomy of the convolutions, that portion of the brain in which we believe, with good reason, exist the structures through whose action the higher functions are manifested, and consequently through the diseases of which insanity is manifested.

The histological components of the brain are :—

1. Nerve fibres.
2. Nerve cells.
3. Bloodvessels.
4. Lymph spaces and canals.
5. Interstitial matter.
6. The cerebro-spinal fluid.

Nerve Fibres.

There are three great systems of fibres in the cortical substance and the immediately subjacent white matter. The first consists of nerve fibres continuous with the peduncular fibres of the spinal cord, which, after passing through the pons and crura cerebri to the optic thalami and corpora striata, radiate outwards to the cortical substance, where, breaking up into bundles, they form what Meynert terms the "projection system." The second consists of certain bundles of fibres which run parallel to the plane of the convolution, and consequently at right angles to the divergent or radiating fibres, forming the first link of Meynert's "association system;" and, third, the commissural fibres, forming the second link.

When we view by a low power a section of a healthy convolution cut transversely to its length, we see the fibres arranged in a fan or fountain-like manner, springing from the central white stem, and diverging outwards to the periphery. As it makes its way from within outwards, each fibre becomes finer and finer; this attenuation has been ascribed by certain authors to dichotomous subdivision, and by others to the giving off of fibres to the horizontal layers. The former opinion is untenable; the latter is open to grave doubt. I am inclined to agree with Cleland that each fibre maintains an individual continuity, and that it terminates, on reaching the outermost layer, by projecting itself horizontally into the superficial structures.

You are all acquainted with the appearance of striation presented to the naked eye on section of the gray matter, i.e., that it is divided into alternating layers of darker or lighter colour. This appearance is due to a system of horizontal fibres which traverse the length of the cortical substance, and to the presence or absence of cells and nuclei in the several layers. The demonstration of these horizontal fibres is much more difficult than that of the divergent, and, unluckily, specimens showing them do not maintain the demonstration for any length of time, in consequence of chemical changes in the necessary hardening agent; and it is perhaps in consequence of this that considerable difference of opinion exists as to their exact anatomy. Lockhart Clarke describes the fibres of the central white stem as, in addition to the radiating fibres, throwing off others which "curve inwards from opposite sides to form arches along with some of the gray layers." This he calls the "arciform system" of fibres. Cleland differs from this opinion; he holds that the horizontal fibres are entirely independent, "that the vertical and horizontal fibres cross one another with the utmost clearness, and without the slightest tendency to pass one into the other." From close examination of a large number of sections, it appears to me that the horizontal fibres immediately external to the gray matter are not derived as described by Clarke, and that those in the actual substance of the cortex hold no immediate connexion with the radiating bundles. It must be remembered that Cleland is most distinct and definite in his opinion expressed in 1860, and that his opinion has not been assailed by Clarke. He is moreover sustained by Arndt to some degree, and Gerlach's observations may be read as favourable. By study of the diagrams of Cleland

12

and Arndt [1] you will gather that there is more than one band of horizontal fibres, that in addition to various individual fibres there are at least three broad striæ—an inner, traversing the convolution where the gray matter meets the white; a middle, thin, in the actual cortical substance; and an outer separating or rather inter-current between the layers containing the largest cells. Outside, there are a few fibres following the course of some of the external layers. [2] From whatever source these fibres may be derived, it is evident that an associating system exists, by which the convolutions maintain very intimate connexions.

Time will not serve for close inquiry into the arrangement of the third great system of cerebral nerve fibres—the commissural fibres of the corpus callosum—I will only remind you that " by them the corresponding and identical regions of the cortex of the two opposite cerebral hemispheres are united " (Meynert).

In the convolutions nerve fibre is reduced to its smallest dimensions. When taken from a recently-killed animal, it presents itself as a simple tube, but shortly after death, or after the application of water, it shows a double contour (Bennett). Certain histologists consider that this is simply a post-mortem appearance, and that during life the nerve fibre or tube consists of a simple sheath or neurilemma containing a homogeneous plasm. But as all post-mortem appearances must result from some variety of structure during life, it can hardly be contended that the manifest differences between the contents of the neurilemma, or primitive sheath, as disclosed by the microscope after death, are entirely artificial. These are doubtless due to coagulation; but the fact that they coagulate in different manners shows that they are of different consistence and different structure. It is reduced almost to demonstration that in life nerve fibre is of a semi-fluid consistence, but this is true of many other structures, muscle for instance; still no one will presume to say that the striation and non-striation of the two kinds of muscle is a mere post-mortem appearance. The fact that we have varieties presented by nerve fibre—I allude to the medullated fibre of the cerebro-spinal axis and the non-medullated fibre of the sympathetic system—shows that the differentiation is not alone caused by post-mortem changes, but by actual structural difference, and of this we have further confirmation by the variety of behaviour of the contents of the neurilemma under chemical agency.

As I have said, the nerve fibre consists of a primitive sheath or neurilemma. This invests two materials or structures, the presence of which causes the appearance of a double contour when the fibre is viewed either in its long or short axis. The outer line is the primitive sheath, the two inner lines the axis cylinder of Purkinje. Between the sheath and the axis cylinder there exists a substance, oleo-albuminous in its chemical composition, to which is due the

---

[1] Schultz's Archives of Microscopic Anatomy, 1867, vol. iii. p. 476.
[2] Bucknill and Tuke's Psychological Medicine, p. 613.

white appearance of nerve fibre; this is the white substance of Schwann. It can be easily seen to ooze out of the torn end of a fibre (Bennett), and it has a strong tendency to assume special forms. It is not colourable by any known chemical agent, in which it differs from the sheaths and the structure it surrounds, viz., the axis cylinder. In recent specimens this axis cylinder presents itself as a gray streak in the length of the fibre; when subjected to colouring agents, as a strongly-tinted band. When speaking of the neurilemma or sheath, I should have said that it is liable to become varicose or ampullated by pressure; this is not the case with the axis cylinder, which always maintains its rod or band-like form. It can be seen projecting beyond the torn or cut end of a fibre parted into strands, which indicates distinctly that it is of firmer and more fully organized consistence than the white substance of Schwann.

It should be mentioned that Gerlach holds that the horizontal fibres of the convolutions are non-medullated. I am not aware that this opinion has been accepted generally, except in the case of the outer thin bands or individual fibres which Cleland believes to be non-medullated.

Nerve Cells.

Lying amongst and between the fibres of the cortical substance, generally following their direction and closely associated with them, we have those all-important structures, the nerve cells; the organs whose special function is, we believe, to receive, generate, or transmit nervous influences.

Your attention has been already directed to the concentric lamination of the cortical substance (in connexion with the horizontal fibres), which was stated to be due, along with these fibres, to the presence, in greater or less number, of cells, and their arrangement and distribution.

This diagram illustrates the cortical layers of the occipital lobe coloured with hæmatoxilin. The preparation from which it is drawn was made from the occipital lobe of a healthy young man who had been suddenly killed by a railway accident. You will see at a glance that the gray matter is divided into six concentric layers. I also present a preparation coloured with carmine, kindly lent me by Dr L. Clarke, which illustrates the arrangement and distribution of the cells as described by him. You will first notice that all the cells follow the direction of the radiation of the fibres, their apices being outward; and, secondly, that they vary in shape, size, and number in various layers. The outer layer is thin, and contains few cells; in it are the ultimate fibrils of cell poles and of nerve fibres. The second is somewhat thicker, and holds numerous small cells and nuclei; the third is pale and thin, with few vesicular elements; the fourth also thin, but darkly coloured, and containing a large number of bigger cells than in the superjacent layers; the fifth is pale and thin, with comparatively few cells; the sixth is

B

dark, and contains also large cells of about equal size with those of the fourth layer. This layer gradually merges into the seventh, which is white; it is, in fact, the thickest layer of horizontal fibres. (Lockhart Clarke, *vide* Maudsley's Anatomy and Physiology of Mind, p. 60, 2d edition.)

In a question involving such minute anatomical details, it is easy to believe that a considerable difference of opinion exists as to the exact number of layers; but in the present state of our knowledge it is needless for me to enter upon the discussion,—suffice it to say, that the general facts are admitted on all hands.

The nerve cells of the hemispheres have been described as pyramidal, oval, pyriform, and triangular, in shape. I am strongly inclined to agree with the opinion that they are more or less either spindle-shaped in the deeper layers, and pyramidal in the outer, and that the varieties described are due to the direction in which they have been cut in thin sections. The pyramidal cells are truncated at the base, the spindle cells maintain the spindle shape at both extremities. You will see how easy it would be to vary the shape of these cells apparently, if they were presented to the eye in transverse or oblique sections. Besides these large cellular structures, we have smaller ones which also differ in shape. Opinions differ as to whether these should be regarded as really cells, or as nuclei of the connective tissue; but there seems to be little doubt that they belong to the former. In the cerebellum, where they form a very important layer, they have been seen in intimate relationship with the large cells; and in the convolutions of the brain proper they exhibit many connexions which lead to the conclusion that their nature is of the higher order.

Lockhart Clarke states very definitely that the cells of the occipital lobe are almost uniform in size. This certainly does not coincide with the observations of Dr Cleland, Herbert Major, or myself, for we all have preparations of healthy brains in which we can show large cells in this lobe. The largest cells that I know of are in the convolutions on either side of the fissure of Rolando at the vertex—in the ascending frontal and ascending parietal convolutions. In this position, the cells of the fourth layer are very large. The nerve cell, as viewed under the highest available powers, consists of a finely-granular and fibrillated body, marked faintly with transverse striæ, a large nucleus, and a distinct nucleolus or nucleoli (Arndt, Wundt). Bennett asserts that they possess " a delicate cell wall, outside which is occasionally seen a layer of nucleated areolar tissue," and in so doing he stands alone. Although certainly not possessing a wall, the cerebral cell is encapsulated or enclosed in a space separating it from the surrounding tissues. The body of the cell consists of a protoplasmic substance, "similar to, though not identical with, pure protogon." The nucleus is more granular than the body, and the nucleoli are apparently homogeneous in structure.

The smaller or truly pyramidal cells, as shown on section, give off each one apical and two basal processes, but these latter are in reality more numerous. The former either runs outwards to the surface of the convolution, and ends like the terminal nerve fibre by inclining horizontally, or it is prolonged into one of the radiating fibres ; the latter, springing from the angles of the base, are connected with both the radiating and horizontal fibres ; and further, they are, according to Rindfleisch, Boll, and Wundt, connected with one another. The larger or more truly spindle-shaped cells have, in addition to the apical and angular basal processes, a large process, the axis cylinder process, which runs inwards and is connected with the fibres of the main white stem, and forms four to ten processes which spring from their sides. These large cells seem to have very intimate relations with the horizontal fibres by means of these poles, certain of which pass directly into them ; and the cells themselves are figured by Cleland as departing from the usual direction—that of the plane of the radiating fibres—and following that of the horizontal band.

The Germans are in advance of British histologists in demonstrating the connexion established between cell and cell. In this diagram, copied from Boll, is shown the method of association : as the process recedes from the cell it becomes fibrillar or penicillated, and the fine fibrils actually join one another ; thus maintaining an anastomosis not only between cells of the same layers, but also with those of the superjacent and underlying ones.

As we go along, I am desirous of pointing out the extreme intimacy of relations existing between the various regions of the brain by means of its various structures. Thus, we see cell communicating with cell by means of their processes ; by means of the horizontal system of fibres, a direct connexion is maintained throughout the hemispherical ganglia between one part and another, and by the radiating fibres an association is kept up with the internal ganglion and the spinal cord. But remark this, that fibre-communication is maintained by individual fibres, which hold no relation one to the other except through the medium of cell processes. We have no evidence of fibre communicating directly with fibre, or even no certain proof that one series of fibres communicates directly with the others; in fact, all anatomical demonstration goes towards the establishment of individuality and isolation of fibre, the processes of the cells being the connecting links. Thus, the apical processes seem to become in certain instances connected with the fibres as they go to the periphery, and the basal processes loop with the horizontal fibres, and also by means of their recurrent poles with those of the central white stem.

If we accept the theory that these cells possess the vital property of generating, receiving, and transmitting nervous influences, it is not difficult to understand that morbid conditions of their structure must affect most seriously the whole economy ; and further, that whatever tends towards the destruction of individual conductivity of fibre must be equally prejudicial. I hope to be able to show you that pathological changes can be demonstrated not only in the brains of the chronic insane, but in recent and acute cases.

In next lecture, we will discuss the blood distribution of the brain, its lymphatic system, the nature of its connective tissue, and the more recent facts and theories which have been elucidated by physiological research.

LECTURE II.

Blood Supply.

We now pass to the important and interesting study of the blood supply of the convolutions, taking it up at the point where it is left in systematic books on anatomy. Such works state generally, that the minute branches of the anterior, middle, and posterior cerebral arteries, after ramifying in the pia mater, pass into the brain substance; but we will now inquire in what manner and form they pass in and terminate.

This diagram but feebly represents the appearances shown by the brilliant preparation which stands on the table. It is one made by Dr Carter from the brain of a rabbit, into the arteries of which he injected a coloured solution of gelatine. The portion of this specimen, which is figured in the diagram,[1] magnified ✕ 100, is the bottom of a sulcus. The eye will appreciate at a glance the fact, that there is a most manifest difference between the quantity of vessels supplied to the gray and the white matter; and closer observation will show that this is due to their size, length, and distribution.

This preparation (one of many hundreds that Dr Carter has made) demonstrates two anatomical facts:—

1st, That there are three distinct series of cerebral capillaries.

2d, That the ultimate distribution of these capillaries varies in the gray and white matter.

You will observe the transverse section of an artery at the bottom of the sulcus, from which spring branches to the right and left, which course round the periphery of the convolution external to its substance. From these smaller branches are given off, which enter the brain substance at right angles to the plane of the convolution. These arteries are various in size and distribution, and may be divided into the *large* or *medullary arteries*, and the *medium* and *small cortical* arteries; the first supplying the white substance, the two latter the gray. The large or medullary arteries run right through the cortical substance, rarely, if ever, supplying to it more than one small twig; as soon as they reach the white substance, they divide dichotomously to the right and left in the plane of the inner horizontal fibres—of course at right angles to the

[1] See Bucknill and Tuke's Psychological Medicine, Plate IX., Fig. 1, p. 614.

direction of the parent vessel. In the white matter they form a comparatively coarse capillary meshwork; this is due to the large divergent branches being connected by short annectant vessels at somewhat wide intervals. The mesh may be stated to be five times longer than it is broad. This appears to be the general arrangement of the capillaries of the white substance.

It will next be observed that there are medium-sized vessels, about twice as numerous as the large. These run directly (giving off a few twigs in their course) to the *inner* layers of gray matter, where they branch off in every direction and at *acute* angles, and then we have the smaller and most numerous arteries, whose duty seems to be the nutrition of the outer layers of gray matter. The cortical arterioles anastomose most freely with each other, but *they are terminal vessels in the gray matter.* Their anastomosis is procured by a much less definite arrangement than that of the white matter arterioles, as they simply divide and subdivide, casting off branchlets upwards and outwards—in fact, in every direction but downwards. I think the arrangement of these three sets of vessels can best be compared to that of a close forest. We have first the straight undergrowth, the branchlets of which mix with the branches of the half-grown trees; the branches of which again are overtopped by the lower ones of the giants of the forest.

The anastomosis of the cerebral arteries is most intimate. In the pia mater we have one anastomotic system, which is continued as the branches from its vessels penetrate the actual brain substance. In the pia mater lobe is connected with lobe, in the gray matter convolution with convolution, in the medullary substance all the inner aspects of the brain maintain mutual relations. The close anastomosis which is so evident to the naked eye between the external arteries becomes increased as they reach their ultimate ramification—in fact, it may be said that there is no ultimate ramification of a cerebral artery, in that it meets its fellow somewhere or other, until they conjointly merge into the venous capillary. Where the artery ends and the vein begins, we are in the same doubt as in other organs of the body.

Along with Dr M'Kendrick I have endeavoured, by simultaneous injection of artery and vein, under equal pressure, to elucidate this point in anatomy. At the present moment we do not feel justified in generalizing on the results of our experiments, but I trust that next session something definite may be laid before you.

I ask you to remark for one moment the difference between the *direction* of the supplies of blood to the white and gray matter. To the white it is by ever-recurring right angles:—the original vessel of the pia mater sends off its branch at a right angle, which again at the same degree transmits its branches to the brain substance, which, thirdly, decussate in a similar manner, which, fourthly, communicate with each other by annectant capillaries at an angle of 90°. The arteries of the gray matter, on the other hand, proceed, after the first rectangular branching, directly to the parts to be supplied, throwing off branches at acute angles, and in every possible direction.

This explains the immensely greater blood supply to the gray matter as compared with the white, which may be roughly stated to be as five to one. Although physicists are agreed that direction does not retard the amount or rapidity of the flow of fluids in *rigid* tubes, I am not aware that they have come to the same conclusion as regards elastic ones. The hypothesis may therefore be thrown out, that the constant recurrence of right angles on the arteries of the white matter *may* have an influence in modifying the amount of blood supplied to that part. However that may be, we have an absolute certainty of the greater vascularity of the gray over the medullary matter; and, arguing from all analogy, what better argument could we have that its functions are of the higher order?

It has been for long taught that the vessels which enter the convolutions are accompanied or supported by the pia mater, or, to speak more accurately, by projection of its inner flocculent surface, the *tomentum cerebri*. But it appears to me, that this teaching is more the result of argument by analogy than of actual demonstration. In fact, in none of the British standard works on anatomy is the pia mater, which is said to accompany the vessels, enumerated amongst their coats, or spoken of as a sheath, the reader being left to the belief that the outer fibrous coat of the artery and the cerebral substance are in direct apposition. In Germany it seems to be admitted that there is a coat outside the outer fibrous coat. Rindfleisch says "that although the arteries of the brain are usually said to enter naked, it would be inexcusable in any one familiar with the morbid anatomy of that organ to overlook the sheath of connective tissue, which, however slender, surrounds its arteries."

I present to you now a series of preparations which seem to me to demonstrate the existence of a coat intervening between the outer fibrous coat and the cerebral substance. Three of these are taken from the brain of one person who had been the subject of senile insanity, and who died from exhaustion and old age. On post-mortem examination no disease of heart or kidneys was found to exist. The pia mater was thickened and adherent. The other morbid appearances need not be alluded to, as they have no bearing on the point at issue. Specimens of the brain were hardened in chromic acid, and the sections were cleared with turpentine and set up in Canada balsam diluted with turpentine; glycerine or acetic acid were not used. The first of the three is a transverse section of the lower part of the medulla, near the decussation of the pyramids. In the anterior median fissure a vessel of considerable size can be seen entering, accompanied by pia mater; at a certain point, the vessel, the muscular coats of which are much thickened, is cut off short, but the empty sheath runs inwards for a considerable distance, becoming thinner and thinner, less and less fibrous in appearance, until it is lost in the substance of the medulla as a fine translucent hyaline membrane, shrivelled longitudinally, and therefore apparently somewhat fibroid. In the second section,[1] which is taken from one of the convolutions bounding the fissure of Rolando at the vertex, a similar vessel and membrane are visible, the latter distinctly pronounced in the degree to which it had arrived at its innermost portion in the first section, viz., the hyaline condition. It is $\frac{1}{150}$th of an inch thick, and is traceable almost through the cortical substance and beyond the inner cut end of the vessel. In the third preparation, taken from the same convolution, there exists in a longitudinal section of a vascular canal, in which no trace of ordinary vascular tissue can be detected, a very fine, somewhat puckered, translucent membrane, non-fibrous and non-fenestrated, filling up the space almost from side to side and from end to end; its edges are well defined; it is not coloured by carmine. I could have produced many more preparations illustrative of this condition of the vessels, but prefer to take my stand on those made from this individual case for the following reasons:—1st, That the sections were not prepared with glycerine or acetic acid. 2d, That the same appearances are noticeable in different portions of the same brain. 3d, That the external pia mater was much thickened. 4th, Because the thickened conditions of the other coats of the vessels was such as to render their demonstration peculiarly easy; and therefore enabling the eye to determine their relation.

I present to you one other preparation. It has been taken from the fresh brain, and is unprepared by any chemical agent, except cold water. The vessel was carefully dissected out from the centrum ovale, and cleaned with camel's-hair brushes, water only being used. Around it you will see a loosely enveloping sheath apart from the outer fibrous coat, forming at the bifurcations a triangular sac.[2]

Judging from my own experience, and from the study of that of others, I can see no reason why we should not regard this membrane as the extension inwards of the pia mater, and therefore as a normal investment of the minute cerebral arteries. I would not have dwelt so long on this pia-matral investment were it not that the fact of its absence or presence has very important bearings on the next question at issue, viz., the cerebral lymphatics.

When we consider the immense vascularity of the brain, the multiplicity and activity of its functions, its consequent constantly called-for supply of nutrient plasm, and its consequent rapid waste of tissue, we will at once see of what paramount importance it is to the study of its pathology that we should arrive at some conclusion

---

[1] See Bucknill and Tuke's Psychological Medicine (3d edition), Plate IX., Fig. 3.
[2] Loc. cit., Fig. 2.

as to the means it possesses for relieving itself of superabundant material and the products of waste. I believe that on this one point hinges the explanation of the causation of the majority of cases of insanity, and that it has important bearings on the pathology of coma, convulsions, and other allied cerebral symptoms. At present, I will confine myself to the anatomical question of the locality of the cerebral lymphatics, as its physiological and pathological bearings fall to be considered under the head of the influences inducing and producing insanity.

Fohmann and Arnold first described the pia mater as possessing an extensive system of lymphatics—an opinion which has been endorsed by subsequent observers. These lymphatics are said to be in direct communication with what is called the *great epicerebral lymph space* or cavity, which is stated to lie below the pia mater, between that membrane and the brain. It is said to have been proved that such a communication exists, as the epicerebral lymph space has been filled by injection from the true lymphatics of the pia mater. Up to the time of Robin, no communication had been shown between the interior of the brain and this space. He, in 1853, subsequently in 1855, and in full detail in 1859, demonstrated that this communication was established by means of lymphatics which followed in the course of the bloodvessels; which, shortly after, His believed he succeeded in injecting from the lymphatics of the neck; according to Eberth, however, he has lately resiled from this position. The subject has been well worked at since by Obersteiner, Roth, and Boll, from whose papers and those of the authors already mentioned this description is mainly taken. These lymph spaces not only follow the course of the vessels, but they actually surround them, and were at first in consequence called perivascular *canals*. The term canal is now abandoned, and perivascular *space* adopted, for reasons which will presently be mentioned. In most sections of cerebral tissue cut transversely to the direction of its vessels, their cut ends will be seen surrounded by a clear ring of unoccupied space, and occasionally fine trabeculæ will be found running between the outer apparent coat and the brain substance. This clear space is, to some extent, due to retraction of the cerebral tissues, consequent on the use of hardening agents; but in diseased brains, as we will see further on in the course, it is also the result of hyperæmic dilatation during life. Considerable confusion has arisen from these clear rings, *which are not the lymph canals of the brain.* I trust I have demonstrated to your satisfaction that there exists a sacculated coat outside the outer fibrous coat, which separates it from the brain substance; for between it, the so-called hyaline membrane or sheath of pia mater, and the outer fibrous coat is the space which is held to be a lymph space by the German authors. This sheath can be traced for a long distance loosely enveloping the vessels until (according to certain German observers) it debouches along with them into the epicerebral cavity by funnel-shaped mouths. According to Boll, the inner surface of the sheath is quite smooth, the outer rough or shaggy (zottig), holding connective-tissue, or, as they are called, after their first describer, Deiter's cells, their rough or shaggy projections communicating with the cerebral substance as if for the support of the sheath. It is believed that the superabundant plasm of the blood exudes into these lymph spaces by a process of exosmosis in the case of the arteries, and that the waste products are got rid of in a similar manner by the veins, at least to a considerable extent; for Obersteiner goes still further, for he holds that by certain "spur-like" processes there is maintained a lymphatic communication between spaces surrounding each nerve-cell and the perivascular lymph spaces.[1] This most careful observer states that he has actually thrown injections into spaces surrounding the cells; and, moreover, that he has seen lymph corpuscles in these spaces. In my own preparations of diseased brains, I have frequently found large clear spaces around the cells, and also in healthy animals, such as the ape; but I am not able to substantiate his position further than this. Obersteiner's theory is a most attractive one, for it gives us a passage by which the products of the waste of very active structures can be got rid of, and passed into the waste-pipe. I am not aware, however, of any confirmation of his observations.

I need not remind you that the brain is not singular in possessing perivascular lymphatics; for, according to Macgillivray, they exist around certain of the vessels of the liver; according to E. B. Kiber and Tomsa, around those of the spleen; and according to Dr Goodfellow, a similar relation exists in the cornea.

It is obvious of what vast importance it is to the pathology of the brain to establish the existence of a lymphatic system in it. In this country, little or nothing has been done in this direction, but in Germany it is held to be an absolutely proved anatomical fact. I propose, during the coming summer, to endeavour to inject these spaces from the lymphatics of the neck; for it appears to me that the method adopted by German authors, with the exception of His, who, as I have said, has resiled from the position, is not entirely satisfactory; they use the "einstick methode," which consists of driving injections under the pia mater, or into the substance of the brain, and allowing them to flow in every and any direction. Doubtless this method has served to demonstrate perivascular lymphatics in other organs, and as the results obtained from it are fully accepted by such authorities as Stricker, Rindfleisch, and Von Recklinghausen, we are justified in generalizing on the probable pathological results of cerebral conditions interfering with the action of these lymph spaces.

But I must confess to not understanding how these lymph spaces can communicate by funnel-like openings with the epicerebral lymph space which is said to lie between the pia mater and the brain

[1] "Wien Stzb. d. k. Akad. Wissener," Bd. LXI., 1 Abth., Jan. 1871.

C

matter. This to my idea is simply impossible. I have never seen
any indication of the existence of an epicerebral space, and when
we reflect that the surface of the pia mater is flocculent as it
is applied to the convolutions, it would seem that its attachment
must be intimate. On the other hand, I have frequently seen
in morbid brains strong indications that there is a space between
the two fibrous layers of pia mater; that, in fact, that membrane
should be described as consisting of three layers,—an outer and
inner fibrous, and a middle or vascular and possibly lymphatic
layer. In the preparation, of which this diagram is a represen-
tation, such an arrangement can be seen most distinctly; and
in this we see the funnel-like opening debouching actually into
the middle, not into a space subjacent to the inner layer. If these
spaces surrounding the vessels, between their outer fibrous coat and
their sheath, are lymphatics, which I think is highly probable, they
must communicate with the actual lymph passages of the pia mater.
How can we believe in a space corresponding to the so-called
epicerebral space in connexion with the spinal cord, where the
pia mater is so very closely adherent to its substance, dipping into
it, and holding it firmly together? It is only in disease that it
becomes separated, and that its supporting influence on the entering arteries becomes impaired. Demonstration, analogy, and
argument are all opposed to the existence of an epicerebral lymph
space, as existing between the naked convolution and the pia mater;
on the contrary, they converge to the establishment of the proper
lymphatics of that membrane being the receptacles and conduits
for the removal of the products of waste.

Interstitial Matter.

Maintaining and supporting the nerve fibres and cells and the
bloodvessels in their position, holding all together as a matrix, we
have a plasm called the neuroglia, or nerve glue or cement. There
exists a difference of opinion as to whether this neuroglia is to be
regarded as the connective tissue of the brain or a nervous structure. It would take up too much time to enter here fully upon this
discussion; I will only say that the weight of argument bears
upon the theory of Virchow and Kölliker, that the neuroglia is
connective tissue. It consists of an intensely fine reticulated
tissue, holding a clear homogeneous protoplasm (which after death
becomes slightly molecular) and numerous nucleated bodies. In
certain positions, and more especially in diseased brains, connective tissue cells can be demonstrated; they are spider-like in
form, and are called, after Deiter, Deiter's cells—why, I can hardly
understand, for their presence was indicated by Virchow in the
first edition of his Cellular Pathology. Scattered in considerable
quantity throughout the substance of the neuroglia are nucleated
bodies, the nuclei of the neuroglia; these are finely granular,

with nucleoli, and present the appearance shown in this diagram, which is an enlarged copy of a plate in Paget's Pathology.[1]

I will not dwell longer on this brain element, all-important as it is; I will only ask you to regard it as the jelly-like matrix which holds *in situ* all the other more highly organized structures. Like all connective tissues, it plays a most important part in morbid processes, being subject to changes in quantity and quality, and to pathological alterations of its cells and nuclei.

As the cerebro-spinal fluid will claim especial attention when we come to the pathological conditions of the brain, I only allude to it now as an important item amongst the brain elements.

And now, Gentlemen, I trust you will not think that I have been too prolix on the anatomical part of my subject. Detail was inevitable, for if there had not been laid before you facts connected with the normal histology of the brain, it would have been impossible to demonstrate the changes which take place in the several tissues. Before leaving the subject, allow me to lay before you Bucknill's theorem: *that we have a right to presuppose that, in the brain as in the other organs of the body, the normal exercise of function is dependent on a perfect maintenance of the anatomical relations of the component structures.* And this is most especially true in the case of the brain, for two reasons. The first is, the extremely intricate relations which its various structures maintain one with the other; we have three systems of fibres, the radiating, the horizontal, and the commissural, keeping up communication between convolution and convolution, between lobe and lobe, and between hemisphere and hemisphere—we have cell communicating with cell, cell connected with fibre, and vessel with vessel;[2] in a word, it is the organ of the body which, in its histological associations, forms the most perfect whole. It is a perfect dual organ. The second reason is that it is, so to speak, self-contained. It cannot, like the lungs or liver, cast any of its functions on other organs; it can gain no relief in disease from vicarious aid; it must do its own work, rid itself of its effete matter, and of the products of injury or disease, and provide within itself for the resumption of functions, the exercise of which have become impaired from whatever cause. The bearing which these facts have on the pathological position is—1st, *That circumscribed lesions may affect the whole encephalon temporarily or permanently;* and, 2dly, *That permanent loss of function of one hemisphere may be supplemented by the action of the opposite hemisphere.*

This latter consideration leads us naturally to the question: What are the functions of the hemispherical ganglia, and what light has modern research thrown upon it?

[1] Page 113.
[2] It is not maintained that the anastomosis of vessels extends to a direct communication between the hemispheres; only as regards the lobes of each hemisphere.

We were all educated in the faith, founded on the observations of such physiologists as Majendie, Longet, and Matteucci, that the cortical substance was the organ through which the highest manifestations of intelligence were alone produced, that it possessed no sensibility, and in no manner subserved the inferior functions of nerve force. Brown-Séquard even could not evolve from his extensive experiments that in it were seats of definitely localized function. But within the last few years a new school has sprung up, the apostles of which are Broca, Hughlings Jackson, Fritsch, and Hitzig, Nothnagel, Gudden, and last, though by no means least, David Ferrier. The theory of this school is, that in the cortical substance of the gray matter of certain convolutions, and in certain parts of convolutions, exist psycho-motor centres —that is to say, that in certain circumscribed tracts ideas are stimulated which prompt, or produce, or excite certain definite and distinct sets of muscles to action.

Fritsch and Hitzig were the first to demonstrate the fact that the gray matter *is* susceptible to irritation—the only wonder is that this fact lay so long in obscurity. They also showed that irritation gave rise to muscular action on the side of the body opposite to that of the hemisphere irritated, and they likewise were able to produce certain of their movements definitely. These physiologists, however, stopped short at this point. But Ferrier, independently of them, and with other objects in view, took up the investigation, and by a long series of experiments, conducted in the highest scientific spirit, has evolved results which will make his name historical in the annals of physiology.

As the accuracy of Ferrier's observations has been lately impugned by a former pupil of Brown-Séquard, Dr Eugène Dupuy, I think it advisable to narrate shortly what I and other Fellows of this College have seen of them.

Early in August of last year I accompanied Dr Ferrier to the Brown Institute at Lambeth, in order to see him prepare two animals, a monkey and a cat, for demonstration to an audience, amongst whom were Virchow, Liebreich, and Burdon Sanderson. With the rapid skill of an experienced surgeon, he reflected the scalp and removed the calvarium of both the animals, arresting the hæmorrhage by pledgets of cotton-wool steeped in perchloride of iron (the animals were fully under the influence of chloroform). The dura mater was removed, and the subjects were left for two hours to overcome the shock of the operation. The monkey very soon recovered, and showed no indications of *malaise;* on the contrary, within an hour and a half of the operation he partook of milk and bread-and-butter from the hand of the operator, jumped on his shoulder, and played with him. The cat could not be trusted.

At the appointed hour a considerable audience was assembled, and Dr Ferrier, without any previous experimentation, proceeded to his demonstration; and in this wise: Holding the monkey in his left arm, utterly unrestrained, he said to the assembled crowd of observers:—" Gentlemen, I am going to touch with the electrode this portion of the brain of this monkey, on which he will open his mouth and protrude his tongue: "—the electrode was applied, the mouth opened, and the tongue was protruded. Next, Dr Ferrier indicated the spots which, when stimulated, would be followed by extension or retraction of the paws on the opposite side of the body, retraction of the ear, rotation of the head, movements of the eyes, clutching movements of the paws, and so on. Each prediction was made with the utmost confidence, and with good reason; for in every instance the result followed with the utmost exactitude, until stimulation became complicated, and epilepsy was produced; but a short interval of rest sufficed for the renewal of the demonstration. Dr Dupuy states that he has failed in obtaining the same results as Ferrier: this must be due to some imperfection in his method. For instance, he says that he has never succeeded in procuring opening of the mouth and protrusion of the tongue: now this very result Ferrier produced with the utmost certainty more than a dozen times on the occasion alluded to. Although by no means at one with Ferrier as to the explanation of the phenomena he has demonstrated, I can vouch for the absolute accuracy of his statements *as to fact* in every particular.

One important fact elucidated by these experiments is, that functional hyperæmia is produced on the application of the stimulus. I ask you to bear this in mind, as bearing on future remarks.

I must refer you to the original paper for full details, and will enumerate now only such as have special bearing on our subject. These are:—

1. The anterior portions of the cerebral hemispheres are the chief centres of voluntary motion, and the active outward manifestation of intelligence.

2. The individual convolutions are separate and distinct centres; and in certain definite groups are localized the centres for the various movements of the eyelids, the face, the mouth, tongue, ear, neck, hand, and foot.

3. Action of the hemisphere is generally crossed; but certain movements of the mouth, tongue, and neck, are bilaterally co-ordinated from each cerebral hemisphere.

4. The cerebellum is the co-ordinating centre for the muscles of the eyeball. Each separate lobule is a distinct centre for special alteration of the optic axes. The cerebellum is the organ on the integrity of which depends the maintenance of the equilibrium of the body.

5. That Dr Hughlings Jackson's views as to the causation of epilepsies and chorea by discharging lesions of the different centres in the cerebral hemispheres is correct.[1]

Nothnagel's system of experimentation, which consists in the production of limited lesions of the gray matter by means of the injection of a concentrated solution of chromic acid, tends to support Ferrier's views; as it is found that the destruction of the ascending parietal convolution results in loss of muscular sense in the fore-leg of the opposite side of the body. Nothnagel also records other confirmatory observations.

Gudden likewise reports some most important results having a similar tendency. He has removed portions of the hemispheres from newly-born rabbits, and from observations made months after the operation, comes to the conclusion that "there is reason to locate the organic conditions of voluntary movements in the cortical substance of the brain, and that they are situated in the frontal lobe."

As opposed to the views of this school of physiologists, we have the experiments and opinions of Dupuy. That he has failed in obtaining the same results as Ferrier is no argument, as, as I have already said, those of the latter are too regular and sequential to admit of doubt; but he asserts that the same results can be obtained by irritating different parts of the gray matter, and even the stimulation of the dura mater produces cross action in the fore-legs. This observer believes that the electrical current must be propagated to the base of the brain in order to excite movement, and he has repeated the experiment of Flourens of removing the entire cerebral hemispheres without any resulting loss of muscular action; and he has further supported his theory of the propagation of stimulus, by showing that a galvanoscopic frog is thrown into a state of contraction when its nerve touches a part of the cerebral mass far from the point of excitation.[1]

This latter observation is really of no value as against the theory of Ferrier; for it is well known that during the passage of nerve force along a nerve, the natural electro-motive force of the nerve is diminished, it is affected by what physiologists term a negative variation. This negative variation is sufficient to stimulate a second nerve laid upon it, in the same way as we see in Matteucci's induced contraction. In the same manner it is evident that, if a small area on the surface of the brain be stimulated by electrodes, other portions of the brain containing nerve fibre communicating with this area will suffer a negative variation sufficient to stimulate the nerve of the galvanoscopic frog. The contractions observed by Dupuy do not show a diffusion of the electrical stimulus, but simply negative variation of other portions of the brain.

The theory of the existence of psycho-motor centres in the brain is by no means definitely settled. My own observations on the morbid histology of the brains of the insane do not appear to me to support it; for, as we shall see further on in this course, *bilateral*

---

[1] Burdon Sanderson's experiments in connexion with this subject have been published since the delivery of this lecture.—*Proceedings of the Royal Society,* 1874.

disease of the cortical substance can exist in the regions indicated by Ferrier as psycho-motor centres without loss or impairment of muscular power, and the cerebellum can be utterly disorganized to the extent of five-sixths of its substance without any aberration of the motility of the eyes, or loss of co-ordination of muscular action. We will postpone, however, the consideration of this question until there have been laid before you the nature and distribution of the cerebral lesions.

It is needless, Gentlemen, to recapitulate the various theories which have been advanced as to the *modus operandi* of cells and fibre in the evolution of psychical phenomena; for in this matter no leading *fact* has been experimentally demonstrated. We know nothing of the direction of electrical currents in the brain; we only know that such exist. Mr Dewar and Dr M'Kendrick lately directed your attention to the all-important fact that the influence of light could be reduced to absolute demonstration in the case of the special sense of sight, and that modifications of these influences could be measured with perfect certainty. We cannot yet see whither this system of experimentation is to lead us. If it be possible to demonstrate the mechanical operativeness of the influences which act on the organs of special sense, why may we not hope that the activity of the forces which we call psychical may yet be recorded by the galvanometer? Are emotion, volition, and actuation one whit less palpable in their immediate effect on the individual than light and sound? We all know the immediate physical effect of emotions; we blush, we pale, we grow hot or cold under their influence, the organism standing in the position of an elaborate galvanometer. When we think of the vast strides made in the science of physics within the last few years, when we regard the concentration of experimentation and ratiocination on the conduct of metals under peculiar conditions, and the vast outcome which has resulted, may we not hope that, when the physicist and the physiologist are more thoroughly merged in one common observer than they are at present, physiology will be able to assert itself as a perfect science? Without a perfect physiology we cannot have a perfect pathology. Till this is established, we can only go on working for future ages, in the hope that our labour may tend to support the observations of the coming man.

LECTURE III.

Influences inducing and producing Brain Disease.

Having in the last two lectures considered the normal condition of the brain elements, their relation to each other, and the later theories of their action and function, we will proceed to the next stage by studying the various somatic conditions which act on them so as to produce solution of continuity. These conditions may be classed as follows:—

I. *Idio-encephalic* disease, or disease affecting primarily the cerebral tissues.

II. *Evolutional* conditions of the body concurrently implicating the brain.

III. *Morbid* conditions of the body concurrently affecting the brain.

In a paper published some years ago,[1] I employed the term Idiophrenic to express the first class of conditions; but Dr Bucknill has improved upon it, by using the more comprehensive word Idioencephalic, which includes not only diseases of the brain substance proper, but also those commencing in the membranes and bones. In the same paper the pathogenetic classification was more diffuse; but the one now before you contains no departure from its principle, merely a modification resulting from extended experience and the study of the opinions of others. Pardon me, if I state now that in this course of lectures other modifications will be made—indeed, in one or two instances actual change of written opinions will be expressed; but I trust that vacillation will not be laid to my charge when resiling from positions which have been found to be untenable. Every man lives through many phases of opinion and belief as work and experience bear upon him, and I have yet to learn that he is the less wise man who openly recants or modifies theories expressed in his scientific juvenescence.

Were we to take the first class alone, and to conjoin the second and third, we would have Van der Kolk's simple classification of insanity; for by "Idiopathic" he expresses primary changes in the nerve structure, and by "Sympathetic" all changes caused by abnormalities of the system conditioning the brain. I adhere to the first position of this great teacher, but consider that the second may be with propriety subdivided, as it takes no cognizance of the normal abnormalities (if I may be allowed the paradox) concurrent with development and decay, apart from the consequences of bodily disease.

I propose considering the Idio-encephalic conditions productive of insanity under four heads—1st, Traumatic; 2d, Adventitious products; 3d, Over-excitation of the brain; and, 4th, Defective organization. The evolutional conditions, under five heads: 1st, Pubescence; 2d, Pregnancy; 3d, The puerperal state; 4th, The climacteric period; and, 5th, The senile period: and the morbid conditions of the body which concurrently affect the brain under three heads, 1st, The Diathetic; 2d, The Toxic; and, 3d, The Metastatic.

Gentlemen, If this course of lectures extended twice its prescribed length—if I had to deliver twelve instead of six lectures—my difficulties would be lessened by one-half; for were the time more extended I should be able to discuss in more thorough sequence the pathogenesis, the symptomatology, the morbid anatomy, and the treatment of insanity. As it is, the long hiatus of twelve months between the present and the next course necessitates the concentration of our energies on one particular point, notwithstanding that there are others of equal interest. As I have already

[1] Journal of Mental Science, vol. xvi. p. 196.

indicated, it is to pathology that we will mainly direct our attention, and if this be done to the prejudice of symptomatology and therapeutics, it is only for the moment, or, to speak more accurately, for the year: for next session these are the points which I propose, amongst others, to descant on. Of course they must be alluded to immediately, but only as they bear upon the question of pathology.

The terms mania, delusion, monomania, melancholia, dementia, dipsomania, and others, which have been used to indicate forms of *mental disease*, will be employed to indicate merely *mental symptoms incident upon certain conditions of the body*. It is needless now to enter upon a discussion as to the propriety of this step, more especially as it must be taken up when we come to speak of the principles of classification; suffice it now to say, that to my mind these terms have no reference to the bodily condition, history, age, diathesis, or idiosyncrasy of the individual, and that two or more of these terms may apply to the same individual within a very short space of time, and therefore have no right to be spoken of as diseases. In the whole of psychiatric medicine there are only some three or four conditions which have claims to the position of pathological entities; for instance, general paresis, idiocy, cretinism, syphilitic, and perhaps rheumatic and gouty insanity; and you will oblige me by bearing in mind, that when these terms are employed in association with those merely symptomatical, it is not right to reduce them to the same level.

You will notice that amongst the idio-encephalic conditions, I have placed last on the list that group which apparently should be considered first, viz., the congenital defects; and for this reason, that certain of these defects are so closely associated with the developmental conditions of insanity, as to render it highly desirable that their study should be undertaken in as close juxtaposition as possible. I would also point out that inflammations of the meninges and brain substance are not taken into consideration under this class, as I can adduce no evidence of their being idiopathically affected.

By *Traumatic causes*, I desire to indicate simply those produced by direct violence to the head. Attempts have been made to associate with them sunstroke, but there is not the slightest ground for the union.

It is believed, and I think rightly, that one very common cause of idiocy is injury sustained by the head prior to birth or during the process of delivery. Attempts at the procuration of abortion, and the violence to which the infant's head is subjected during its passage through a contracted outlet, or by the application of mechanical aids, may produce such traumatic lesion of the brain substance as to cause arrest of its development, and consequent idiocy. That so few children brought into the world by means of the forceps become idiots, is another instance of the wonderful power of recuperation possessed by the brain: but it is well worthy of remark, that

D

statistics show, as far as statistics can illustrate such a complicated question, that there is less danger to the future psychical health of the being so assisted into the world, than by leaving it to the unaided efforts of nature to procure its passage through an abnormally contracted pelvis.

It must have come within the range of the experience of most present to observe how very frequently extensive injuries of the brain and calvaria are unaccompanied by permanent psychical symptoms: on the other hand, it must equally have come under your observation how apparently slight injuries have resulted most seriously to mental health. I need not recapitulate here cases which are, so to speak, historic; reference to the works of Pott, Morand, Hecker, Ogle, Ricker, Gross, Chevalier, Le Gros Clarke, and others is sufficient. Within my own experience, I can refer to a case in which I removed considerable portions of the left frontal bone of a boy whose forehead had been wounded by the kick of a horse; to another, of a naval officer whose skull-cap was cracked diagonally by a shell, almost from one ear to the other; and to a third, in which a bullet shattered the superior portion of the right parietal bone, carrying with it meninges and brain matter—all of whom made good recoveries without any subsequent serious loss of intellectual power. In the first case, the boy got well under adverse circumstances, without a single bad symptom; he was never comatose, feverish, or inclined to vomit. In the two latter cases, very prolonged coma followed the injury, but when consciousness returned the mental faculties were almost as good as they had ever been, with the exception that concentration of attention on any particular or difficult subject for any length of time produced intense headache, that the temper became somewhat irritable, that the memory was slightly impaired, and that one glass of wine caused singing of the ears and a sensation of numbness about the head. In all these three cases the discharge was very considerable from one side of the brain.

It has also been my lot to meet with several cases in which simple concussion was followed by very serious consequences to the mental faculties of the patient; also with one case in which simple depressed fracture of the occipital bone resulted in a condition similar to general paresis. Of the cases of simple concussion, two were caused by railway accident; the immediate symptoms were not severe, headache and sleeplessness being the principal ones; but on these gradually supervened inaptitude for business, loss of memory, unfounded suspicion, and a tendency to undue taking of stimulants. These two cases were almost identical in their symptoms; they both slowly recovered; many months, however, lapsing before the cure was complete. Judging from published cases and my own experience, it would seem that injuries of the head in which the skull has been fractured and hernia cerebri has occurred, are less likely to result in permanent insanity than those in which no lesion of the scalp has existed. It cannot be said that this depends on the greater immediate mortality of the two kinds of injury—at least, if we exclude from the category the slighter cases of concussion, the symptoms of which pass off quickly,—for, taking cases of fracture and of injury without fracture relatively as regards the severity of their causation, it will be found that they are very nearly equal as to their results in immediate death, but that they are not relatively equal as to the subsequent permanence of their results to psychical health. The reason of this is not far to seek. In the compound fracture, the extravasation of blood and the products of inflammation find free exit : in the simple fracture we have a permanent cause of irritation, either from depression, from spiculæ, or from the so-called *vis medicatrix naturæ*, which kills by the production of plasms which can get no vent.

Through the kindness of Dr Heron Watson, I was enabled to examine microscopically the brain of a young man who last winter was killed by being struck on the forehead by the buffer of a railway engine ; he was rendered comatose, and died in eighteen hours after the accident. The calvaria and scalp were uninjured ; dissection showed nothing but a very small extravasation of blood in the medulla ; various portions of the brain were hardened in chromic acid, cut, and set up in the usual way. Sections of the occipital lobe showed thin extravasations of blood between the pia mater and the brain substance, due to rupture of the minute arterioles passing inwards from that membrane. Specimens from this case are now on the table, from which it will be gathered that these apoplexies were so thin as to be unrecognisable by the naked eye—in fact, that they could not have been observed except *in situ*, as the stripping off of the pia mater would have removed them along with it. The vessels of the subjacent brain substance were engorged with blood, and numerous small apoplectic clots, varying in size from $\frac{1}{50}$ to $\frac{1}{25}$ of an inch in their greatest diameter, were found in the external layers of gray matter. In the specimens before you, these are represented by *lacunæ* or open spaces, from which the coagulated blood has escaped during the process of setting up. This case presents, to my mind, points of important pathological interest ; it is true it stands by itself; but it was a commonplace accident, and the appearances presented by naked-eye examination of the tissues were of the commonplace unsatisfactory nature. It is the microscopic apoplexies in the brain substance and those between it and the pia mater which demand especial attention and interest—an interest which extends to injuries beyond the encephalon. I allude to injuries of the spinal cord from concussion. Such lesions of continuity as rupture of the vessels must of course impair the nutrition of the subjacent nerve tissue, and as a consequence are followed by paralysis of function. In the brain, coma is the result—that is to say, paralysis of its function ; and unless the injury is so far circumscribed that the function of the affected territory can be quickly taken up by corresponding portions of the opposite hemisphere, death is the result from the depressent effects on the general system. But the white matter of the spinal cord does not possess the same power as the brain of easting on the corresponding organs of the opposite side the exercise of functions temporarily or permanently impaired in one hemisphere. Rupture, accordingly, of the small nutrient arteries within the layers of the pia mater or at their entrance into the nervous substance must produce impairment or obliteration of the function of the neighbouring parts of the cord, which, in consequence, either undergo further degenerations, greater or less in degree according to the severity of the causating lesion, or gradually recover as the close anastomosis of the capillaries supplements the work of the injured nutrient branches. In this consists, I believe, the pathology of railway spine.

The psychical symptoms which are produced by traumatic injuries of the head are various in kind. Profound dementia of a permanent character is occasionally although rarely met with ; this probably is due to numerous small apoplexies in the outer layers of gray matter destroying the relation of the cells. As has been already said, symptoms allied to general paresis have been known to ensue;—this from chronic irritation. Epilepsy with its concurrent psychical symptoms is likewise a frequent sequence. But the majority of such cases are symptomatized by change of character and the exhibition of morbid impulse. I will only allude to two : the first that of an officer whose action in the field went far to influence the result of one of the most important Crimean battles ; he was promoted on the field : some years after, when in a position of trust, he met with an accident by which he was precipitated from an inconsiderable height on to his head, which fell on loose sand ; he was stunned at the time, but soon after evinced a change in his *morale*, gradually became a hopeless dipsomaniac, and eventually committed suicide. Dipsomania, it may be mentioned in passing, is a very common form of insanity under this condition. The second is reported by Dr Charles Skae, of Ayr, and will be found fully detailed in the January number of the *Journal of Mental Science* of this year. The following is a short summary :—A collier was struck on the head by a mass of coal, which produced a depressed fracture of the skull three inches above the left eye : he gradually became altered in character, epileptic, maniacal, and dangerous, having attempted to kill his wife. He was removed to the Ayr Asylum, where Dr Charles Skae caused the operation of trephining to be performed over the seat of the depression : the operation was successful in every way ; the epilepsy ceased, also the maniacal outbursts, and in a few weeks the patient became perfectly sane. This is an important and suggestive case, not only in its bearings on pathology, but also on practice.

I am loath to part from this interesting subject, but time compels me to push on to others as important.

Adventitious Products.

By the term adventitious products, it is meant to indicate all forms of tumours of the brain, of exostoses of the calvaria, and of bony deposits on the membranes. Doubts have been expressed by certain authors, as to whether tumours of the brain are really productive of insanity, their objections being founded on the fact, that in certain cases tumours have existed which have not been accompanied by abnormal psychical symptoms, without reference, however, to the position of these growths, or to certain points in cerebral physiology, which might have assisted them to explain the position. But I think no one present doubts the fact, that tumours of various descriptions are absolutely productive of insanity, notwithstanding the variety of symptoms by which their presence is evidenced, and the occasional non-confirmation of diagnosis by postmortem examination. The frequency of occurrence of tumour of the brain amongst the insane is, judging from the experience of many observers, very varied. Writing in 1872, Dr Clouston says :[1] "At the Carlisle Asylum we have performed 214 autopsies, and there have been 6 cases of tumours of the brain, which is at the rate of 28 per 1000." In the Somerset County Asylum cases of tumours of the brain were at the rate of 16 per 1000. Dr Sutherland mentions that he found them at the rate of 20 per 1000 ; Bucknill and Tuke at 2·5. Fischer had not one case out of 318, and I am in the same position ; for, out of 400 autopsies of insane persons of which I have records, there is not one in which tumours of the brain are recorded. I have, however, on three occasions met with gliomata in the sane.

Dr Boyd has compared the relative frequency of the occurrence of tumour in the sane and in the insane, employing the data afforded him by his experience of 1039 post-mortem examinations in the St Marylebone Infirmary, and 875 at the Somerset County Lunatic Asylum. In the former institution he found 22 cases, being at the rate of 18·3 per 1000, in which no mental derangement was observed, and in the latter 14, or at the rate of 16 per 1000. The total number of cases of tumour of the brain which Dr Boyd found amounted to 38, of whom 17 showed symptoms of insanity, which is curiously in accordance with Calmeil's observation, that one-half of cases of tumour of the brain are accompanied by more or less aberration of intellect. This estimate, however, is doubtless too high, if we include adventitious growths generally.

Of all the varieties of brain tumours, carcinoma appears the most productive of perversion of psychical function, and this is probably due to the great rapidity of its production. In the more slowly formed tumours the nutrition of the brain is not so much interfered with, and the severity of local irritation is diminished, and further, time is given for the portions of the opposite hemisphere to take

[1] Journal of Mental Science, vol. xviii. p. 153.

up the function of the implicated regions. It very frequently happens that tumours at the base of the brain are not accompanied by any indications of insanity ; on the other hand, those of slow growth occasionally produce changes in the relations of the superior convolutions of such a serious character as to necessarily implicate the mental faculties ; for by pressure " they cause portions of the gray and white substances of the convolutions to pass through small openings in the dura mater, to embed themselves in the cranium, and so form true herniæ of the brain." I quote from a most interesting paper by Dr Clouston, which is the best monograph on the subject extant;[1] it has only one fault, that it is not yet finished. The following are the conclusions he has come to, as regards the pathological influences exercised by tumours growing in the brain. He says that they have three distinct effects on the brain structure.

" 1st, They create an irritation tending to ramollissement in the nerve substance, with which they are in contact from their first appearance. 2d, They cause pressure on distant parts, which in its turn, causes an alteration of the structure and nutrition. 3d, They set up progressive disease, and degeneration of certain parts of the nerve structure, the true nature of which is as yet not very well known ; but it seems to be in some way directly connected with the essential nature and constitution of all sorts of nerve substance, whether cells or fibres. Its results pathologically are—an increase of the connective tissue in the form of granules, and enlargement and thickening of the coats of the bloodvessels ; but all these seem to be secondary changes."

Some diversity exists between the observations of the various physicians who have written on this subject as to details, but they all seem pretty well agreed as to the leading characteristics of the mental phenomena. These appear to be irritability and loss of self-control, perverted morale, gradually increasing dementia ; and as in the case of the traumatic influence there is an absence of delusion, and a tendency towards morbid impulse. In many, symptoms allied to those of general paresis present themselves. Dr Clouston states "that irritability and loss of self-control, and a change of disposition, are the first mental symptoms of those tumours of the brain which directly produce morbid psychosis; that the depression present seems to result from the patient's knowledge of his probable incurability, and is therefore natural : that a blunting of the whole of the mental faculties soon comes on, and gradually passes into coma : that there is a distinct and strong analogy between the symptoms, mental and bodily, produced by such large tumours, and those of general paralysis, which is the type of progressive degenerative diseases of the nervous system, inasmuch as it affects the brain and sympathetic ganglia and retina : that such cases would seem to hold an intermediate place, so far as mental symptoms are

[1] Loc. cit.

concerned, between acute inflammation of the cortical substance and blood poisonings on the one hand, and hereditary insanity on the other, the mental characteristics of the three being represented by delirium, irritability, and delusion respectively."

It is worthy of remark, that a general resemblance exists between the psychical symptoms of those mentally affected as a consequence of tumour and of traumatic injury of the brain. In both there is, as a rule, an absence of delusion, a more pronounced tendency towards morbid impulse than is observable in any other form of insanity, with the exception, perhaps, of that associated with epilepsy, and to a gradual decadence of the general system—all pointing, however, to one common cause, perverted nutrition, with its sequential lesions.

Remarks on syphilomata and tubercular deposits are deferred until we consider the diathesis with which they are associated.

But apart from these manifest tumours, we have a very important series of conditions implicating the connective tissue of the brain, which can only be demonstrated under the microscope. I allude to various forms of sclerosis of the neuroglia, to colloid degeneration of its nuclei, and to the morbid bodies which we call amyloid.

Under the head of Over-Excitation of the Brain, we will consider what are generally called the *Moral Causes of Insanity*. In systematic works and asylum reports the causes of insanity are usually divided into two great classes—the physical and the moral; the former being referred to conditions of the body, the latter to conditions of the mind. We will shortly weigh the arguments *pro* and *con* for such a differentiation, and inquire how far and to what extent the so-called moral is in fact a physical cause.

It is not for me to remind you, that emotion acts as an inducing morbific influence on organs remote from the brain. Under certain circumstances the trophesial or nutritive influence of the central organ is exercised with extreme rapidity on remote structures by, we believe, its power over the vasomotor nerves. Within the confines of health we see this influence manifested in the blush of shame and the pallor of fear, the goose-skin of terror, and the diuresis and diaphoresis of an emotion, compounded of anticipation, doubt, anxiety, and fear. Beyond the pale of health we have, amongst many other changes caused by the influence of emotion, the perversion and arrest of milk in the nursing woman, and the diarrhœa and paralysis of fright. I was once witness of the latter phenomenon in a very marked degree. A soldier belonging to a distinguished regiment, with which I once had the honour of serving in a temporary official capacity, was tried for striking his superior officer during time of war. He was a strong muscular man, in good health, and had proved himself a brave soldier in action, although a bad one in time of peace. He was tried by court-martial. The regiment was formed in square to hear the finding of the court and its confirmation, and to witness the carrying out of the sentence: the verdict was guilty, and the sentence that "Private A. B. shall be shot to death by musketry." Private A. B. dropped as if the sentence had been actually carried into effect: he had not fainted, but had lost completely the use of his legs: he was supported till the rest of the proceedings were detailed. These were, that, in consideration of certain circumstances, the sentence was commuted to forty lashes and penal servitude. As soon as A. B. realized the fact that his life was to be spared, he regained all his faculties, walked to the triangles, which were presently rigged, stripped, and took his punishment like a man, well laid on as it was, without a groan, and marched to hospital when it was over.

Proceeding higher in the pathological scale, we occasionally meet with cases in which more permanent and serious bodily disease is induced by sudden emotion. I will instance only one case—one, however, which doubtless will bring to the memory of each and all of you instances of a similar nature. A lady, who had been confined of her fourth child, all her previous confinements having been of the easiest nature and shortest duration, was, when all was going on quietly and comfortably, suddenly informed that a beloved sister had died in childbed: she immediately had a rigor, and in three hours symptoms of peritonitis set in with great intensity. No symptoms of puerperal insanity showed themselves; she was comatose for many days, was brought to death's door, but eventually pulled through, and is now the happy mother of children, another having been born with the utmost ease since the confinement spoken of.

Emotion may be the cause of death. I do not speak of those cases of sudden death from joy or sorrow which are on record, as we have not their clinical history, without which they can have no pathological value. You will find, in a work called "Old New Zealand, by a Pakehah Maori," a description of rapid death from chagrin, as occurring amongst the natives of those islands. I was witness of two such cases, but will relate only one, as they were almost identical. A strong healthy young man, belonging to a well-known native family, became, without any known cause, disgusted with life: he had no indication of any bodily disease, refused food, but not liquids, and did not attempt to commit suicide. He took to his bed, and in less than a week after the first manifest symptom was found there dead. Such cases are not uncommon amongst the South Sea Islanders.

It would be beyond my province to endeavour to explain here the *modus operandi* of emotion in producing morbid changes of remote organs, more especially as the subject has been carefully investigated by many authors, more especially by Dr D. H. Tuke, whose interesting book on the "Influence of the Mind upon the Body" is well worthy of study. It is for us to seek an explanation of disease of the brain, and the physical changes which take place in consequence of prolonged or undue emotional action.

I have already stated that Ferrier's experiments show that the stimulus of Faradization causes a determination of blood, a functional hyperæmia in the part of the brain submitted to its influence, and that certain phenomena are consequent thereupon. Of course it is and may always be incapable of absolute demonstration, that the same condition exists in parts stimulated by the exercise of their intellectual, volitional, or emotional functions; we can only argue that such is the case from extraneous evidence. Every man in this hall has experienced the sensation of fulness and throbbing of the temples and forehead, accompanied by coldness of the feet and legs, after prolonged and absorbing study, followed by sleeplessness and headache. No one chooses the hour immediately succeeding dinner for active thought, or, if he does, he does it to the detriment of the process of digestion. It is needless here to produce the various witnesses at our call in support of the opinion, that when the blood-supply is increased, activity of the brain is produced, and that it is diminished, *pari passu*, under opposite circumstances. The analogies of other organs afford correlative evidence.

But functional hyperæmia, which is in itself normal so long as it is requisite, becomes productive of abnormal results if it is continued longer than the occasion requires, as it causes unnecessary activity of the supplied tissues, and consequently unnecessary prolongation of the exercise of their function, and prevents a condition necessary for repair—that is to say, the anæmia of sleep. Again, blood too suddenly and in too great quantity determined to certain portions of the brain by the stimulus of emotion, may produce explosive symptoms by immediate over-stimulation of the cells. We have thus two abnormal conditions of blood-supply—the first in degree chronic, the second in degree acute. The chronic, far the more common, is the immediate cause of loss of sleep, with its manifold sequential symptoms; the second, comparatively rare, causing immediate over-excitation of the cells, and perversion, temporary or permanent, of their function.

The investigations of Durham, Hughlings Jackson, and Hammond, have definitely determined that during sleep the supply of blood to the brain is materially diminished; that, in fact, during sleep the brain becomes anæmic, and the rapidity of the flow is materially modified. Against this may be instanced the stories told of active brain-work carried on and completed during sleep; of problems solved, and sequential trains of thought worked out in dreamland: these are mere statements, unsupported by aught but supposed experience. Those who have believed that they had undergone such experiences were either not asleep, or the results they supposed they had obtained occurred during the minutes or moments of wakening, when the blood was acting on cells already stored with the factors of sequential thought. Such stories hold no position in scientific pathology.

If you will each and all analyze the cases of insanity which you

E

have ascribed to undue emotion, I believe you will find that a very small proportion indeed are referrible to it as the *immediate* cause: I mean, that deliberation will show that a considerable interval has occurred between the emotion and the first indication of absolute mental alienation—an interval during which sleep has been absent in consequence of continued hyperæmia, primarily caused by emotional influence. I have appealed to the experience of many physicians who have devoted their lives to the study of mental disease, as to the relative proportion of immediate and sequential cases which have come under their observation. I will refer you to the experience of one such physician, Dr John Smith, of this College. His experience, which extends over half a century and many thousand patients, and which is supported by accurate written records and a most retentive memory, can instance no well-marked case in which insanity has been proximately caused by emotion: he has known of several in which the allied symptoms of epilepsy and catalepsy have been produced by sudden fright, but the search of records and the taxation of memory can produce no cases in which mental aberration, more or less persistent, has resulted. I speak without the book, but within the mark, when I say that you will find that cases the immediate results of emotion stand in no higher ratio than 1 to the 10,000 of those in which mental symptoms show themselves, as the result of loss of sleep induced primarily by emotion. Within my own experience, I have never met with an instance. Of course you will understand that such complications as pregnancy, the puerperal state, or lactation, are not involved in the present discussion.

The rarity of the immediate causation of insanity by abnormal explosion of emotion is all the more curious, when we remember the comparative frequency of its morbid influence on remote organs. It would appear that great provision is made for the over-exercise of function of the cerebral tissues, and that they are less liable to sudden morbid change than the parts from which they receive and to which they convey influences.

Although the limits of cerebral health are broad, they can be overstepped by the withdrawal of the condition of sleep, during which the brain tissues are rested and repaired. Disappointment from reversed fortune; the carking care of the speculator in love or business, the anxiety of the fond husband, father, or mother, watching at the sickbed; the retrospection of the mourner on the days that are no more; the weariness of anticipation; doubt and hope deferred; the despair caused by the austerity of certain forms of religious beliefs, their plenitude of damnation and their straitness of salvation—are but a few of the so-called moral causes which induce loss of sleep. *They are not moral causes, they are absolutely physical.* They bear but slight analogy to the causes of degeneration of the tissues of other organs, inasmuch as they take their rise in the normal exercise of the functions of the cells, which, passing beyond normality, results in certain physiological modifications of blood-supply necessary to health : in this consists the difference between emotion and such morbific influences as cold, heat, and septic or other poisons.

Over-exercise of the intellectual function is not by any means such a prolific cause of insanity as undue emotion. It is not work but *worry* that kills the brain. When both are combined the result is often rapid. Take the case of Sir Walter Scott : how hard he worked, and how happily he worked, in the earlier part of his literary career ; but when he was compelled to labour and tax a brain, already overtaxed by worry and anxiety, how fearfully rapid was the breakdown ! A converse case was that of Whewell, who, in the calm atmosphere of the academy, worked to the last day of his life, with but slight modification of his intellectual power, notwithstanding that evidence was found after death of brain atrophy. The forced intellectual work of Scott was a very different thing from the automatic cerebration of Whewell. The exercise of the imagination to supply under adverse circumstances ever-called-for " copy," bears no relationship to the outcome of a brain trained to systematic ratiocination—a brain from which flowed without compulsion, almost without volition, the results of the stored-up impressions of years : the brain of the one succumbed to work and worry ; the brain of the other retained its intellectual faculties to the last, those tissues only being affected by senile decay, the degeneration of which does not imply loss of thinking power. Scott's cells underwent degeneration from loss of sleep ; Whewell's maintained their continuity from the due recurrence of repose.

In the absence of the anæmia of sleep, the over-stimulated cells give off their outcome in gradually increasing intensity, till the patient goes mad, possibly from the loss of inhibitory power. The primary emotion as a rule, though not always, influences the characters of the excitement, which is generally of short continuance, leaving the patient weak and exhausted. Unless relief is soon found, the changes in the cells are accompanied or followed by lesion of other brain structures, which are productive of most important pathological conditions affecting the general system, which at times result rapidly in death. These will be considered separately : they are only referred to now to remind you that there is no change in cerebral tissue which has not its effect on the body generally.

I trust you will not think that this particular subject has been dwelt on too long, for I believe it is one of the few keys to the general position. It has, moreover, most important bearings on the theory and practice of psychiatric physic.

We now pass to the last of the idio-encephalic conditions productive of insanity.

Defective Organization.

In the wide domain of medicine, we are constantly compelled to ascribe certain trains of diseases to a defect or defects in organization, pathological conditions the nature of which have not been

elucidated by the various scientific instruments at our command—
conditions, however, which are easily recognised by the *mens medica* of the practised observer. We are accustomed to speak of these
as neuroses, diatheses, or tendencies, words implying an instability
of relationship of the constituents of the different systems in which
the trains of diseases are manifested, which renders them peculiarly
liable to abnormal nutritive changes, and consequently to perversions of function. Is this assumption of a general tendency
towards degeneration a cloak for our ignorance? I think not—it is
only an assertion of the amount of knowledge we at present possess.
We can fairly point out the ætiology, the pathogenesis, the symptoms—in a word, the general pathology of these conditions. They
are entities, save in one thing, their histology. We cannot lay our
finger, or rather I should say our eye, on the defective elements.
What is true of the other systems of the body is peculiarly true of the
nervous system. We cannot say in what nerve element or elements
the defects primarily occur. We have no data to go on in ascrib-
ing nervous instability to a deficiency in bulk of a particular constituent, to an impairment or increase of conductivity of nerve fibre,
to a deficiency or over-activity of chemical action of cells. I am not
pleading *ad misericordiam*, when I say that instability of the nervous
system, neurosis, diathesis—call it what you will—is in no worse
plight than the other well-marked morbid general conditions to
which you apply the several terms, in accordance to the various
systems through or by which you regard pathology. It is no disparagement to neuropathy to admit that the nervous system is
liable to conditions of instability which have not, and probably can
never be, submitted to ocular demonstration. But it would be
feeble to cast aside as useless from an imperfect science a broad
generalization, founded on the observation of all, simply because
one link was missing in the chain of evidence.

To enter upon a full description of the natural history of this
constitutional peculiarity is quite uncalled for now, as every one
present is well acquainted practically with the subject. I only ask
you to remember that it is claimed as a very important idio-encephalic condition, productive of insanity. I refer those who desire to
become acquainted with the best writings on this question to the
works of Maudsley, Bucknill, Tuke, and A. Mitchell, where the
various interesting points connected with its heredity, its influence
on the morale and the intellect, and the corporeal frame, are fully
discussed. I am free to confess that here is the debatable land on
which the psychologist and the pathologist meet on the fairest
terms; for they both are equally armed with hypothetical weapons,
neither of them possess the lance of absolute demonstration;—the
only difference being that the psychologist *thinks* he has it, and the
pathologist *knows* he has it not. As one man may be tubercular,
and yet not suffer from tuberculosis, as another may be gouty
without developing what we call gout, so a third may be consti-

tutionally predisposed to nervous instability without ever becoming insane. We specialists await the incidence of manifest symptoms of insanity, as you do the indications of phthisis or podagra, before we assert any absolute pathological rights.

LECTURE IV.

Mr President,—I find that so much time has been occupied in the consideration of the anatomical and physiological part of my subject, and of the idio-encephalic conditions, as to necessitate a somewhat cursory review of the evolutional causes of insanity, that is to say, those which are concurrent with the changes in the development and decay of the body; and also of those conditions which are concurrent with diseases of organs remote from the brain. Were I to enter in detail upon those subjects, it would be impossible to lay before you fully that department to which I have directed special attention—the morbid histology of the brain—and which I am naturally desirous to illustrate. This will be done in the last two lectures.

As has been already said, the evolutional conditions of the body which concurrently affect the brain, require to be studied in close association with the last-named idio-encephalic condition—instability of the nervous system,—and for this reason, that the exciting cause of the insanity occurring at the climacterics (to use the term of the ancients) is so comparatively slight, so common to all mankind, that it almost presupposes an hereditary or congenital defect in the organs implicated.

It may be said that this is true of all forms of insanity—that a man who becomes insane must have been predisposed to the disease; but with this I cannot agree, any more than I could with the statement that the subject of, let us say, chronic catarrhal pneumonia, must have been predisposed to diseases of the respiratory system; or, to take a better instance, that all cases of Bright's disease are dependent on some instability of the constituents of the kidneys. We may grant this to be true of the gouty or contracting, but not of the acute inflammatory, form of these diseases: and in like manner, we have a right to refer insanity connected with the pubescent or climacteric period to nervous instability; whereas we should not be justified in assuming the same of a case of mania incident upon bodily disease or toxic influences. That weakness of constitution may be a factor in the production of all forms of degeneration, whether of the lungs, the kidneys, or the brain, is no doubt true; still, in certain of them, it is by no means an essential one, for they may result from causes non-inherent in the individual. I would not have given expression to this truism, were it not that so many mistakes, implicating the comfort and happiness of individuals and families, are made, in attaching the stigma of a tendency to so-called mental disease.

To the public, madness is madness—no attempt is made to differentiate between madnesses and madnesses : to have had an insane member in a family is to cause the world to regard with suspicion the general psychical health of every member, without regard to the cause or to the kind of the insanity in the afflicted individual. The manifest injustice of this is palpable to the medical profession, as its members know to what various causes and influences insanity is ascribable, apart from hereditary predisposition. Thus we may have, and constantly do have, insanities arising concurrently with morbid conditions of the body as mere accidents in the course of the case, in very much the same manner as delirium or convulsions, and probably due to a similar cause or causes, from which the patient recovers without any permanent influence on his psychical health. The utmost care should be taken, in forming and expressing an opinion, as we are so frequently asked to do, as to the heredity of insanity in a family, to obtain the fullest evidence as to the general pathology of the case or cases of the individuals who have been its subjects.

But, after all, why should we speak of insanity as a stigma on a family, any more than we should of cancer or tuberculosis? Doubtless it is a terrible disease, but in what is it a more terrible disease than consumption? Only in this, that it is a more inconvenient disease. Is it in itself more loathsome than cancer or a dozen other maladies to which flesh is heir? Is it less curable than the majority of the graver forms of disease? Here it certainly has the advantage; for, given two persons, the one with the preliminary symptoms of tubercular phthisis, and the other with the preliminary symptoms of acute mania, *cæteris paribus*, of which would we predicate the greater probability of recovery? I am sure there is not a physician here present who would not prognose in favour of the latter. Society has surrounded insanity with a false metaphysical glamour, with which is compounded a savour of the ludicrous that does not enter into its estimate of what it considers purely bodily disease.

The observation of ages has caused us to accept as an ætiological fact, that numerous and various degenerations commence contemporaneously with modification of nutrition at the evolutional periods of existence ; and it might be fairly asked, Why should mental alienations be classed in accordance with their period of incidence any more than other morbid conditions? To this question it may be answered, that the morbid conditions occurring at these phases of which I now speak are of such a definite nature as almost, if not completely, to warrant their assumption as pathological entities ; that, were the type of other morbid conditions modified in any great degree by age, it would be competent to classify them accordingly, and that the terminology of certain forms of degeneration is influenced by such circumstances. The periods of pubescence and adolescence, the first and second climacterics, and the degenerations of old age, must be regarded as immediate causes of insanity in those constitutionally prone to nervous instability, and also in those who may be placed under such accidental circumstances as may have a depressing effect on nervous energy. The changes in the nutrition of the system which take place at these periods concomitantly influence the psychical condition of the most stable, and in the unstable produce an increase or decrease of functional activity of the brain-cells, which, overstepping the limit of health, leads to the evolution of abnormal emotion, and consequent abnormal actuation, the latter in certain instances tending to aggravate the condition.

Dr Anstie lays great stress on the occurrence of convulsions during teething without apparent adequate cause, as a strong indication of the insane neurosis, which, as the child merges into boyhood, is further evidenced by a tendency to steal, lie, commit acts of cruelty, and manifest cowardice and meanness of disposition. This is one train of symptoms; but there is another, of which a precocious æstheticism is the leading characteristic. It is open to question how far the latter is associated with the tubercular diathesis; in my opinion, the association is close and intimate. Under either of these circumstances, self-pollution comes in as an adjuvant to render the boy nervously unstable. There are high authorities who hold very strong views as to the evil influence exercised on the organism by the dietetic system of the present day : they maintain that in districts where the diet of the child and boy is simple, where the influence of strong food is withheld, the psychical health is better and stronger than in cities where over-nutrition demands of the youth some outcome of over-excited function. The extreme athleticism of the present day is regarded as a vicarious emunctory of over-excited function, actually called for by a too nutritious diet. Although we may not be all at one with these authorities, it cannot be denied that views, founded as these are on comparative observation, are well worthy of the attention of the profession.

Trying as pubescence is to the predisposed boy, he has yet a harder trial to undergo at the period of adolescence. As the virile function asserts itself on the organism, heredity frequently asserts itself. The evil habit of pubescence may have been continued uninfluenced by the self-respect which ought in the man to have overcome it; over-exertion of the brain by study may have deteriorated the vesicular elements beyond their power of recuperation ; or the altered and altering nutrition may specially excite to action the function of the cells of the hemispherical ganglia ; each condition producing a peculiar train of symptoms. Thus, the masturbator becomes peculiar, suspicious, reserved, full of faint fears, and self-accusing; the student melancholic, suicidal, and misanthropical; and the sthenic subject evinces maniacal excitement, irritability, a desire for action—he walks, talks, drinks, smokes, must ever be doing something. But whatever may be the general symptoms of these three sets of patients, they have one common symptom—a perversion or increase of the sexual instinct—the masturbator thinks women are looking at him, that he influences women unconsciously to themselves; the melancholic believes himself impotent; and the sthenic would run amuck amongst women. It is the last class which so often develop dipsomania. Insanity at this period is said to be influenced by sex; it is stated to be of more frequent occurrence amongst young women, and that the symptoms generally partake in a great degree of the character of hysteria. This is certainly not my experience, which may be exceptional—nay, is, if compared with that of most writers on the subject. I grant the hysteria, but not the insanity. It is contended that the evolutional change is more physically marked in the girl than in the boy. I fail to see this as a reason. I view it conversely; that the *absence* of an emunctory in the boy is more likely to make him prone to nervous disturbance than the girl, who has a physical vent. I can only instance my experience that absolute insanity is more frequent in lads, whilst the modified insanity of hysteria is more common amongst girls.

Between the adolescent and the climacteric periods the constitution of the nervous, like the other systems, becomes established, and disturbance is not liable to occur, except from some accidental circumstance apart from evolution. But when the great climacteric is reached, we find a singularly well-marked form of insanity manifesting itself *without any adequate cause* apart from the change of life. In those who have abided by the laws which regulate health, in those who have met with the success and comfort which as a rule accrue from a well-spent youth and middle age, in those whose constitutional proclivities are not towards nervous degeneration, this period is overcome with but slight constitutional disturbance, the middle-aged man or woman simply lapsing into the man or woman of advanced years. But in converse cases, the trophesial influence which grays the head, bows the back, flattens the foot, and modifies the general nutrition, concurrently reduces the vigour of the cerebral constituents. No more definite reason can be assigned for the one than can be for the other series of conditions. It is simply a law of nature that degeneration takes place at certain periods of existence; and as no degenerative change acts equally on all the systems, so each is liable to a predominance of decay, according to circumstances in the individual.

The influence of sex over the frequency of incidence of insanity at the climacteric period is slight. The following is the result of the analysis of 210 recent cases admitted into the Fife and Kinross Asylum:—Of these, 48 presented the well-marked train of symptoms of climacteric insanity—19 out of 86 men, or 22 per cent., and 29 out of 124 women, or 23 per cent. Before commenting further on this point, it may be well to state what these symptoms are. With slight modifications their leading characteristics are a vague fear of impending but undefined evil, a belief that the soul is for ever lost, that the patient has done some wrong to his family, that he is accused of some crime the nature of which he cannot determine, that he or his family is ruined, a strong tendency to suicide for the direct purpose of evading his misery, and, in a large majority of cases, an acknowledgment and recognition on the part of the patient of his own insanity. These are the leading symptoms, modified by idiosyncrasy. The general tone of melancholy pervading the condition may be influenced in the individuals by temperament, as every disease is. Melancholy may be characterized by excitement verging on mania, but still it is a melancholy mania. But what explanation can be offered of the fact that the incidence of these trains of mental symptoms at the climacteric period is equal, or nearly equal, in men and women? In women we have the cessation of a physiological emunctory, the catamenia, to assert as a cause. [But is there not, in man, a cessation, or at least a considerable modification, about this period, of an emunctory? Time will not serve to discuss this point—it is merely thrown out as a suggestion.]

It is unnecessary to do more than allude to senile insanity, as its general pathology is universally admitted. I see that Bucknill and Tuke include it in the idio-encephalic conditions, and I believe I did so myself in a paper written some years ago; but, on the whole, it seems to me now more pathologically correct to refer it to an evolutional condition, which is by no means limited to the nervous system; in fact, that to the morbid condition of the bloodvessels must be ascribed the degeneration of cellular activity and of brain tissue.

Nor will I dwell longer on those other evolutional conditions—the pregnant and the puerperal—interesting and important though they are, than will suffice to discuss the question of their genesis. It has been, and even now is, the custom to speak of the insanities which occur concurrently with pregnancy, follow parturition, and supervene on lactation, under the common designation of puerperal insanity. How erroneous this is must be evident, when we consider how different are the efficient causes and the mental manifestations of these three forms of insanity. The various and profound changes of the whole system which attend pregnancy do not leave the nervous centres unaffected. The most frequent indication of insanity in the fecund woman, is a mere exacerbation of the morbid longings or appetites, and the change of disposition and temper almost invariably accompanying pregnancy. These in the nervously stable subject do not pass beyond a slight and temporary alteration in the morale, which, slight and temporary though it be, would under other circumstances be regarded as morbid; but here we accept it as a part of the physiological process. It is not, however, to be wondered at, that the great modifications of blood supply, of which

F

we have such constant evidence in child-bearing women, should, in those who are strongly predisposed to nervous disease, be productive of actual insanity. How much these changes structurally affect the encephalon may be deduced from the presence of osteophytes or bony plates on the surface of the dura mater and the inner table of the skull; and how much, so to speak, functionally, from congestion and flushings. Nor is it only the quantity, but the quality of the blood which may be a factor. The increase of fibrine and water, and the decrease of albumen, must exercise an influence over the nutrition of cells, and it is possible that the predominance in the individual of one or other of these conditions may determine or condition the nature of the insanity. The most frequent symptoms of the insanity of pregnancy are melancholy and moral perversion : the latter taking the form of dipsomania, the former being accompanied by a suicidal impulse, much less commonly mania of an asthenic type. It is worthy of remark, that psychical disturbance manifests itself most frequently during what are generally considered the critical months of utero-gestation, and that its incidence as to frequency is in inverse ratio to the number of the confinement.

Passing over the mania occasionally occurring during childbed as the head passes the vulva, which is most probably due to temporary congestion consequent on muscular effort, we find a well-marked sequence of mental symptoms following on delivery. This is the true puerperal mania, a name which, when applied to the insanities of pregnancy and lactation, is a direct contradiction in terms. In the large majority of cases of puerperal insanity, the symptoms are those of violent mania—a mania with delirium.

The puerperal maniac has symptoms which, as a rule, cannot be mistaken for any other form of insanity, with perhaps one exception,—mania *a potu ;* but even here there are points of diagnosis which are very prominent. The bodily symptoms are at direct variance with the mental. She is pale, cold, often clammy, with a quick, small, irritable pulse, features pinched, generally weak in the extreme, at times almost collapsed-looking. But withal she is blatantly noisy, incoherent in word and gesture ; she seems to have hallucinations of vision, staring wildly at imaginary objects, seizes on any word spoken by those near her which suggests for a moment a new volume of words, catches at anything or any one about her, picks at the bed-clothes, curses and swears, will not lie in bed, starts up constantly as if vaguely anxious to wander away, and over all there is a characteristic obscenity and lasciviousness. Suicide is often attempted, but in a manner which shows that it is not the result of any direct cerebration ; she may wildly throw herself on the floor, attempt to jump from the window, or draw her cap-strings round her throat, but there is no method about it, it is an impulse, the incentive of which is purely abstract.

This description of course applies to the severest class of cases, but it is taken from the recollection of not a few. Even where the symptoms are not so acute the same tone exists; and the shorter the time of its supervention after delivery, the more acute and marked the mania, and more rapid the recovery.

A careful analysis of a large number of cases of puerperal insanity which had been admitted into the Royal Edinburgh Asylum and the Fife District Asylum showed that, although mania was the most frequent symptom, melancholia was occasionally the indication : but what is of great importance from a pathological point of view, melancholia never appeared within sixteen days after labour, generally manifesting itself in about a month. It further eliminated the fact, that complicated are much more frequently followed by mental symptoms than natural labours; for out of 73 cases of true puerperal mania, 23 (or 30 per cent.) had supervened on abnormal parturition. These two facts taken together lead to the conclusion, that melancholy after labour is not due to the same exciting cause as mania; in a word, that it is not puerperal insanity in the true sense of the word. The various depressing influences of childbed, its various accidents reducing vitality, the sudden return to ordinary physiological conditions, the cessation of normally abnormal physiological conditions, the rapid call for a new focus of nutrition, the translation as it were of the blood to the mammæ, may all be instanced as physical influences liable to act on the brain tissues. The late Sir James Simpson advanced a most seductive theory, that puerperal mania depended on an albuminuric condition, in like manner as puerperal convulsions. As far as my own observations go, and they have been fairly extensive, I have not been able to substantiate this theory; for although I have found albumen in three cases in very slight clouds, still in upwards of 26 no evidence of its presence was obtained. It may be fairly confessed, that the pathology of puerperal insanity is not so definite, or referrible to absolute conditions, as the insanity of pregnancy; but it asserts itself in so strong a manner that we are compelled as physicians to accept it as a special form, although as pathologists we are not able to give such strong reasons for our belief. It is mere hypothesis, still an hypothesis not altogether inadmissible, to suggest that the decadence of those conditions, which had produced such physiological adventitious structures as osteophyte, may be factorial in the induction of insanity. Great difficulties stand in the way of the elucidation of the morbid anatomy of these two forms of derangement. On the other hand, we have in the case of the insanity of pregnancy so many suggestive facts as to actual structural changes, as to lead us to believe that by the study of the causes of the slighter forms of mental disease, we may hope to arrive at a knowledge of those evidenced by more persistent symptoms.

The insanity, when it follows on prolonged lactation, being of a truly anæmic character, falls to be considered under that general condition.

Under evolutional conditions may be fairly comprehended the manifold disturbances of the catamenia—but in a very general manner, inasmuch as they are induced and produced by each and all of the conditions which may be inductive of insanity. It is far too complex a question to enter upon at present, how far the increase, decrease, or absence of the menstrual discharge is influential in the induction of insanity, or how far they are the results of the trophesial influences exercised by the conditions which induce the insanity. This must be relegated to the future, of which I have so often spoken. That the modifications of the discharge are occasional immediate causes of insanity is proved by the fact, that on its regaining its normal amount and quality, the mental symptoms disappear.

A very cursory consideration of the morbid conditions of the system which concurrently affect the brain must suffice at present. But before proceeding to this cursory consideration of individual conditions, I must remark on the absence of abnormal mental manifestations as sequences on certain forms of disease, which, *a priori*, might be supposed to be the most probable immediate causes, but to which in fact they can seldom be referred. I allude in the first place to those chronic diseases which affect the rectum, bladder, uterus, and contiguous parts, such as stone, fistula, cancer of uterus and rectum, stricture with its miserable consequences, and many others which must suggest themselves to you as exercising a most distressing and depressing influence on the nervous system. Of course, we can all remember individual cases in which insanity appeared as a remote consequence on such maladies; but at the first glance, it would be matter of wonder that the instances at our command are so few, were we believers in peripheral irritation as an immediate cause of insanity; and in the second place, I refer to diseases of individual organs, such as the heart, liver, spleen, and kidneys. It might be expected that heart disease would be more frequently inductive of insanity than it is—in fact, it is a question whether it is ever the efficient factor. We are told, to be sure, in certain works on mental disease, that obstructive valvular disease is connected with simple and hypochondriacal melancholy, and that dilatation of the heart is frequent in chronic mania; but these very vague statements carry really no weight with them, as they are unsupported by clinical statistics. Judging from my own experience, I do not think that disease of the heart has any real connexion with insanity as a cause, however much it may have as a result—its early presence in association is a mere coincidence. In like manner, diseases of the liver are, as a whole, innocuous to brain health. We certainly find abnormal livers in insane subjects, but we are almost invariably able to refer the insanity and the liver disease to one common cause. Diseases of the kidneys are decidedly rare amongst the insane, although we have temporary mania or delirium in Bright's disease, probably due to uræmic poisoning.

Much stress has been laid on diseases of the uterus and ovaries, and more especially on tumours of these organs, being the primary factors in the production of insanity. Skae laid down as a special form ovario- or utero-mania ; and Dr Wergt of Illnau has described the various morbid conditions of the female organs of generation found on post-mortem examination, and connects with them mental symptoms. Not having met with any case which on mature deliberation could be ascribed to such diseases with absolute certainty, and only with one or two in which a strong presumption was warrantable, I began to doubt that uterine disorders ever exercised a sympathetic influence, or acted on the nervous centres by peripheral irritation to a degree productive of insanity. On consulting some of our leading gynecologists, I found that their observation tended in the same direction ; that although cases occasionally but rarely presented indications of melancholy, this symptom was more a secondary than a primary one, and was caused by loss of sleep consequent on depressed emotions, anxiety, and pain. The fallacy of advancing uteromania or ovariomania as a special form arises, I believe, out of the clinical observation that, in certain cases, tumours and other deviations from health in the organs of generation, condition to some extent the nature of the mental symptoms in such a form as climacteric insanity. Women who for years have been carrying tumours, when they arrive at the change of life develop aberration of intellect, and not unfrequently the character of their delusion is marked by sexuality and erotomania ; they think they are pregnant, or that they are visited at night by men. It is to this extent only, I believe, that such morbid growths are connected with the insanity ; they may condition, but there is no actual clinical proof of their primary influence on causation by peripheral irritation. There is too much of the element of coincidence—of the *post hoc, propter hoc*—involved to allow us to accept utero- or ovario-mania as a pathological entity, for I have known of several chronic cases of insanity with uterine tumours without any sexual symptom. In like manner, we have a few scattered cases of insanity coexistent with disease of the kidneys, bladder, and spleen, but they all want what an art critic would call breadth of colour conjoined with definiteness of outline, which is so characteristic of the great forms of insanity.

From an ætiological point of view it is of great interest, and I believe of great importance, to note that insanity is seldom, if ever, the immediate *result of diseases of individual organs,* or of peripheral irritation, but that it is most intimately associated with those forms of disease which are the manifestations of a general constitutional instability, such as tuberculosis, rheumatism, gout, or syphilis. This fact, if fact it be, points to the implication of the nervous system and of its central organ in the general impairment of nutrition which produces these morbid conditions, and that as each and all of them may manifest itself in various ways and in various regions, insanity and the brain may be the way and the region in which a section of them do manifest themselves. Thus, without overstepping the limit of health, we have the feebleness or the æstheticism of the tubercular, the irritability of the gouty, and the paralysis of energy of the syphilitic. There is, I believe firmly, an hereditary phthisical brain, and an hereditary gouty brain, and an hereditary syphilitic brain; and further, that through such brains the constitutional instability may be evidenced by insanity without any actual outbreak in other parts of the system.

Passing from the general question, we find that in these three great morbid constitutional conditions, insanity is an occasional concomitant, and the psychical symptoms in each case are well marked and defined. We have to thank Dr Clouston for the first description of what he calls the insanity of tuberculosis. His conclusions have very severely been called in question, but I, for one, maintain a belief in their accuracy and value. The symptoms of their form in no way partake of the character of the spes phthisicorum or euphoria, although in some respects there is a resemblance between the want of fixity of purpose of the uninsane tubercular subject and that of the mentally afflicted one. To quote Dr Clouston's own words, the symptoms may be described as " a mixture of sub-acute mania and dementia, with a great disinclination to exert the intellect; occasional unaccountable little attacks of excitement, and a disinclination for work or even amusement; pervading all is a strong tendency to be suspicious." As to the pathological anatomy of this form of insanity, little can be said. Tubercular deposits on the surface of the brain or in the tissues are as rare in the insane as in the sane adult, and as yet the microscope has contributed no fact bearing upon the immediate question, although it presents a hopeful field of research. There are many very interesting points in connexion with tuberculosis associated with insanity which will be fully discussed when we take up the question of the corporeal symptoms of insanity and the influence it exercises in modifying other forms of disease.

In the insanities which are concurrent with rheumatism and gout, we have as perfect pathological entities as are presented in the whole extent of mental disease. If we were now classifying insanity a difficulty would arise as to where these two forms should be placed; whether they should be held to be diathetic or metastatic. This is especially true in regard to rheumatic insanity, for the mental and bodily phenomena are distinctly vicarious one of the other. The metastasis from the general to the nervous system is symptomatized by choreic movements of the limbs, which disappear contemporaneously with the return of inflammatory swellings of the joints or other indications of the disease in the muscular tissues. But if we accept the general principle laid down in the opening remarks on this section, the metastasis must be regarded as a peculiarity of the general condition, and the brain, and more especially its connective tissues, being predisposed like other structures to the morbid influence of rheumatism, is liable to be, in common with them, the seat of translation.

The general characteristics of gouty insanity are very analogous to those of the rheumatic form. The metastasis and vicariousness of the symptoms are the really important pathological indications. Dr Berthier has given us an excellent paper on this subject. He has, however, omitted, I think, one physical symptom of gout, which is an obfuscation of intellect, or rather a semi-paralysis of the thinking powers in those to whom gout clings, without breaking out into acute inflammation. I refer you to Dr Berthier's careful paper, published in the Annales Medico-Psychologiques, in the year 1869. With regard to syphilis, we may accept the changes in the blood-vessels themselves as the cause of insanity amongst its subjects. This interference with the nutrition of the brain appears to act in two ways : 1st, by producing circumscribed tracts of degeneration and softenings which implicate the health of the whole encephalon; and, 2dly, by a general impairment of its nutrition. I lay before you specimens taken from two syphilitic cases who died insane, in which the remarkable changes in the vessels are very well marked. I am not sufficiently versed in the literature of the disease to know whether the appearances you will see have as yet been described, and I shall feel obliged by an expression of opinion on the subject.[1] With the lately published exhaustive lectures of Dr Broadbent on Syphilis before us, it is unnecessary for me to say more on this subject.

It would likewise be a work of supererogation for me to discuss in full the toxic causes of insanity, for all here present are well acquainted with the various theories of the actions of the poison of alcohol, tobacco, and other, so to speak, articles of diet, and of the toxic influence of the various mineral and vegetable poisons on the system. In the instance of alcohol, the poison which is the prime factor of a special form of brain disease, alcoholismus, acute and chronic, a simple physiological experiment at once demonstrates that its application to an exposed nerve results in immediate cessation or paralysis of its energy ; and as it is fully proved that it is readily absorbed into the circulation, and thus is applied to the central nervous organ, we have little difficulty in accounting for its physiological and pathological actions. In the same way the results of the abuse of many medicinal agents can be explained up to a certain point. But the whole question of brain-poisoning presents far too wide and too often traversed a field for present consideration. It is only introduced here *pro formâ ;* it merely enters its appearance to assert its right to be considered as a cause of insanity. Its pathological histology will be taken into consideration further on in this and the next course.

[1] See " A Case of Syphilitic Insanity "—Journal of Mental Science, Oct. 1874, with plate.

Turning to anæmia as a cause of insanity, we have a somewhat more complex subject to deal with. The defective nourishment of the brain-cells in the state of anæmia must be accepted as the cause of impairment of their functional activity, for they must suffer concomitantly with other structures, and relatively in proportion to their great vital activity, rendering the regions over which they exercise control all the more liable to the morbid processes which are the known results of the modification of blood constitution. The term anæmia is not used here to imply a condition antithetical to hyperæmia ; it does not indicate any mechanical deprivation of blood supply ; on the contrary, the amount of blood, such as it is, is not reduced in quantity, possibly it is increased. The temporary mechanical anæmia which results from extreme cold, for instance, produces its effects rapidly—short delirium and profound sleep, recovery from which is accompanied by intense pain. But the anæmia which qualitatively implicates the constitution of the blood is the anæmia which produces more or less permanent results on cerebral health.

In a region of which the recuperative nervous energy is depressed, the phenomena presented by the corpuscular elements of the blood under abnormal qualitative conditions are all the more likely to manifest themselves. Taking a healthy brain, we might expect to see the corpuscles flowing as quietly and orderly as in the perfectly healthy frog's foot or bat's wing, a demonstration which few of us have seen. Taken, on the other hand, a brain whose activity is depressed in a subject whose blood constituents are under the morbid conditions attendant on anæmia, we might expect with almost perfect certainty that the transit of the corpuscles would present the appearance we have all so frequently seen in an unhealthy frog's foot, the aggregation and rolling motion of the red, and the lagging and wandering of the white corpuscles. It is well known that in anæmia, however it is produced, the red corpuscles have a tendency to coalesce, in consequence of the acquirement of an adhesive viscid quality, causing them to hang together, and so preventing the intercurrence of the white corpuscles. When we take all this into consideration, it is not difficult to see how, even more rapidly than in hyperæmia, a condition of stasis can be established in the cerebral vessels.

The psychical symptoms which are manifested in anæmic insanities are, as a rule, characterized by delirious mania—of not long continuance, but intense while it lasts.

Having thus slightly sketched the leading causating influences of insanity, although by no means having exhausted the category (many of minor importance being of necessity omitted), I will conclude this section of my subject by indicating the great chasm which yawns between the pathogenesis and the morbid anatomy of nervous disease, especially of that portion in which abnormal psychical phenomena are the predominant symptoms, i.e., insanity. Clinical observation shows that we have trains of symptoms, more or less defined, associated with certain efficient causes of psychical disturbance; but in the best defined train of symptoms, let me instance general paresis, we have a wide area of divergence. If there is a pathological entity in connexion with what we call insanity, general paresis is that entity; but mark how very widely divergent are its psychical symptoms. True, in nine cases out of ten of this disease, we have a thorough sequence of psychical symptoms, *but there the odd tenth stands* demanding explanation. Take a more common insanity, drunkenness—ordinary ebriety; why is it that one man differs from another in the manifestation of artificial madness? Take, again, the insanity more rapidly induced by anæsthetic agents, why does one patient quietly subside into dementia, whilst another runs rapidly the course of symptoms which we often note in more permanent insanities,—depression or melancholia, violent mania, and terminative dementia? At this point the pathologist must hold his hand; he falls back on the *interrogatio questionis,* idiosyncrasy, or diathesis. But in thus narrowing himself in his imperfect science he is only relatively narrow to his brother worker in the more accurate science of physics. The physicist is compelled to speak of force, but what is idiosyncrasy but a modification of force in the individual? The physicist stands erect when he shows a change in condition of a bar of iron—a comparatively simple body—under the influence of heat, regarding with condescension, with reserve, and with self-complacent criticism, the work of the man who directs his attention to that somewhat more complicated substance, the human frame. It is not probable that the mechanism of psychical action will ever be projected on a chart, or the molecular and chemical changes which accompany it will ever be submitted to demonstration. Helmholtz tells us that there is a limit to the magnifying powers of the microscope—a limit which has already nearly been reached. But another Helmholtz may arise to whom the most abstruse work of his father in science will be but a mere primer.

LECTURE V.

From what has been said in the last two lectures, it must have been gathered that my intention was to indicate that acute brain disturbances which are evidenced by psychical alienation, are caused by modifications of the quantity or quality of the blood-supply, and that the modifications in quantity are the result of at least five physiologico-pathological processes, which serve to differentiate further between certain of the idio-encephalic and the evolutional and constitutional conditions. In the 1st, we have the primary changes in the cells, which influence secondarily the blood-supply, as in the case of continued loss of sleep and the arrestment of the recuperative quality of the cell; in the 2d, we have changes in the general nutrition of the system, implicating through the medium of the bloodvessels the health of the cell, as in the evolutional conditions, and in certain constitutional conditions; 3d, that in general anæmia the modification in the quality of the blood acts as a prime factor; 4thly, that various toxic agencies produce changes in the nervous tissue which modify the blood-supply; and, 5thly, that changes in the tissues of the vessels, as in atheroma and constitutional syphilis, may produce extensive lesions by mechanical deprivation of blood-supply. Thus, we have in the first condition the cells primarily, and in the other four secondarily affected.

Now in all recent insanities, whether they be symptomatized by acute mania or acute melancholia, the first post-mortem appearance presented to the naked eye is an injected condition of the bloodvessels, and the microscope only serves to render this condition more thoroughly demonstrable — the vessels are found full of blood, distended by it, and fully occupying the cylinders containing them: we have evidence of the existence of a condition of stasis. On the length of time taken to restore the intracranial circulation to its normal condition depends, I believe, the result of the case—recovery, or its continuance in a chronic form. In the case of recovery, a period of convalescence is necessary for the full resumption of the functions of those constituents impaired by the congestion; but if the obstruction be not soon relieved, permanent pathological changes ensue, which may be termed the secondary lesions, and which must be regarded as the causes of chronic insanity.

But how are we to account for this condition of stasis? I confess this is by no means an easy question to answer. No theory of increased mechanical pressure pumping more blood than is normal into the cerebral vessels is admissible, for the exhausted condition of the anæmic maniac precludes such an assumption, even were it warrantable on other grounds: it is to some influence on the muscular coats of the vessels themselves, impairing their controlling function, to which we must look for an explanation of the phenomenon. There are three theories which may be advanced—1st, That the degeneration and impairment of the functions of the cells of the convolutions may exert an influence on the sympathetic centre in the medulla oblongata pointed out by Ludwig, by means of the divergent or radiating fibres, which, as you will remember, are continuous with the peduncular fibres of that organ and of the spinal cord; and that dilatation of the vessel results from loss of their inhibitory function of the muscular fibres. Against this theory it must be stated, that this congestion does not extend to all the vessels of the brain, that as a rule those of the base and of those parts below the level of the horizontal limb of the fissure of Sylvius are not implicated. 2dly, It may be said that the well-known phenomenon of the increased supply of blood to morbid tracts to meet the demand called for by morbid processes offers an explanation; that the myriads of exhausted cells concomitantly depress the contiguous tissues, placing them in the position of actual disease: this appears to be a more hopeful field for speculation. And, 3dly, It is not altogether hypothetical that a change in the blood constituents occurs, the nature of which has not yet been ascertained, but a change in some degree allied to that noted in the frog's foot on irritation of the spinal cord.

It is possible that stasis may result from a combination of these influences, but at the present moment no thoroughly satisfactory explanation of it can be offered ; we only can state the fact that, in all acute insanities, this condition is ever present; and I will now endeavour to point out in what manner congestion of the vessels of the brain is the factor in the production of many of the secondary lesions. But before doing so, I would repeat the observation that it is on *the superior surface of the convolutions that we find the evidence of hyperæmia and its consequences.* On removing the calvarium and dura mater we almost invariably find the arachnoid milky, opalescent, thickened in a greater or less degree, and the pia mater injected most markedly over the area of convolution exposed ; for when we remove the brain from the skull the portions below the line made by the saw are comparatively healthy to all outward appearance ; we find that the focus of disease is at the extreme vertex, at the point where the fissures of Rolando meet, extending backwards and forwards along the line of the longitudinal fissure, and that the manifestations of morbidity fade away laterally and are gradually lost about the level of the horizontal limb of the fissure of Sylvius, and that, except under specific conditions, the base of the brain is to all appearance as healthy in the chronic maniac as in the most mentally stable subject. Simple and self-evident as this pathological fact is, I believe I may claim to have been the first to record it—in the Sixth Annual Report of the Fife District Asylum. What is true of the naked eye appearances is true also of the microscopic ; for, with the exception of the cerebellum, microscopic lesions are more pronounced in parts subjacent to thickened membranes than elsewhere. It is also well to remark here, that opacity of membranes is first traceable in the line of the vessels. With these remarks, we will pass to the factorial influence exercised by stasis of the cerebral vessels in the production of secondary lesions.

It will be remembered that the arteries and veins of the cerebral hemispheres are enclosed in comparatively rigid cylinders of nervous tissue and neuroglia ; that between the brain substance and the proper coat of the vessel we have a fine hyaline membrane surrounding and encapsulating it ; and that by the space, small as it is, between the vessel and hyaline membrane, are conveyed the lymph products to the lymphatics of the pia mater. We have thus a main and overflow, as it were, in one cylinder ; the main, however, is elastic, and is liable to changes in its calibre to such an extent that under their influence it fills completely the retaining cylinder. *When this takes place, the overflow is occluded and rendered useless.* That the artery does in hyperæmia fully occupy the cylinder is proved by the fact, that in subjects in which congestion has been a frequent condition—*e.g.*, epilepsy and general paresis—the cerebral substance is distended and the comparatively rigid cylinder enlarged by pressure from within. But this is to be seen only in old standing cases—in recent ones the vessels appear to be, if you will excuse the paradoxical expression, fuller than they can hold. Under such circumstances, what becomes of those effete or superfluous matters and the exudates produced by hyperæmia and stasis, which should have been carried off from the surrounding tissues, and which may be presupposed to be increased in quantity to the same extent as the blood supplied? They cannot find exit by their normal passages, accordingly the brain substance and the membranes become œdematous, the pia mater becomes displaced by the serum, which, oozing through the tissues, raises it from the convolutions within and without the sac of the arachnoid, fluid becomes arrested, and the ventricles of the organ become dropsical. If relief is not soon obtained the patient dies from some complication of the general system—not unfrequently passive congestion of the lungs. If relief is found, as in the large majority of cases it is, the patient either slowly recovers mental health, or, in consequence of grave and irremediable injury done to the textures of the brain, lapses into one or other of the chronic conditions of lunacy.

It is to the study of these lesions of the brain we will now turn; and if I dwell upon them in pretty full detail, I must ask your indulgence, as one is naturally led to dilate fully on what has occupied many years of one's lifetime as a special field of investigation. That the outcome of this work is but slight and unsatisfactory compared with what might have been expected from the opening up of an almost untrodden field, there is no one more willing to concede than I; but it was necessary to break the crust of the field, and, accordingly, I now lay before you the results of the first turning over of the soil.

In the following remarks will be incorporated the results of the observations of Dr Howden of Montrose, Dr Kesteven of London, Dr Major of Wakefield, Dr Lockhart Clarke, and various Continental authors. This is mentioned now to avoid unnecessary repetition of names.

My own observations have extended over more than a hundred cases of insanity, and 25 subjects who had died without any abnormal psychical manifestations; also the brains of various animals—cats, dogs, moles, rats, monkeys, pigs, etc.—and some fishes, have been subjected to examination, for the purpose of comparison and the establishment of a criterion of health. As to the lower animals, it may be at once said, that in no single instance have any of the morbid appearances about to be described presented themselves—a uniform clearness and definition of structure has been observed.

It was my good fortune in the earlier stages of the investigation to have associated with me Professor Rutherford of King's College.[1] We together examined and reported on thirty cases,[2] and you will all recognise how much must have been lost to the inquiry when this eminent physiologist found himself compelled to withdraw from it. The method of working which we conjointly adopted has been in the main adhered to, at least as far as the preparation of the specimens is concerned, for in other respects it has been much elaborated. It is shortly as follows :—The brain having been removed not later than thirty hours after death, sooner when circumstances permitted, was first submitted to the microscope in the fresh state. Next, portions were taken from the frontal, parietal and occipital lobes, from the cerebellum and corpora striata, pons and medulla oblongata, and from tracts manifestly diseased, and immersed in chromic acid solutions until sufficiently hard for cutting into thin sections ; these sections were then stained with carmine or hæmatoxylin, cleared of water and spirit by turpentine, and set up in Canada balsam. You will remember that in the first lecture it was not contended that this system of preparation showed the exact condition of the brain during life—it was only contended that the modification is so slight as to render the demonstration reliable, in so far as that the relation of constituents is not materially interfered with. That the appearances presented by the sections of diseased brains are not due in any way to the chromic acid or other chemical agents, is proved by the simple fact that the brains of animals or of healthy human subjects do not show any corresponding morbid structures ; on the contrary, that thin sections prepared from them demonstrate the various constituents in regular and unvarying order, in wide contradistinction to the morbid condition.

We will consider now *seriatim* the lesions, 1*st*, of the vessels ; 2*d*, of the cells ; 3*d*, of the nerve fibres ; 4*th*, of the neuroglia ; 5*th*, of the membranes ; 6*th*, of the epithelium ; 7*th*, changes in the amount of cerebro-spinal fluid ; and, 8*th*, certain occasional morbid products. I will first, as it were, simply catalogue the various lesions, and then draw such deductions or advance such hypotheses as may appear tenable.

1*st*, *The Vessels.*—When we examine microscopically a small vessel which has been carefully dissected out from the fresh brain of a subject who has died in a state of acute insanity, it will be found to be full of blood ; when carefully washed with camel's-hair brushes dipped in water and the blood removed, the first thing that attracts the eye of the observer is the presence of large amorphous masses of a dirty yellow material, and a fine granular deposit between the hyaline membrane and the outer fibrous coat in the lymphatic

---

[1] Now Professor of the Institutes of Medicine in the University of Edinburgh.
[2] Edin. Med. Journal, Oct. 1869.

space.[1] The first consists of hæmatoidin—at least of what was lately called hæmatoidin, for the changes in chemical nomenclature are so frequent that one can never make sure whether a name will stand for more than a year or two. Having of late had greater opportunities of examining the brains of subjects who had died in a state of sanity than formerly, I have paid particular attention to the study of these deposits, and have found that hæmatoidin is generally to be found in small quantities on the vessels of most sane subjects, but in a manifestly less degree than in the insane; and that in the non-lunatic subject it appears to depend to some extent on age, and more especially on the nature of the disease which has been the cause of death. In fever cases, in which the insanity of delirium and coma had supervened, it is pretty well marked. But in no instance have I found it in such quantity or in such large masses as in the chronic lunatic. The whole question of hæmatoidin deposits requires careful study, as it points to dissociation of blood constituents under various morbid conditions. The second kind of deposit consists of a very fine molecular material, apparently homogeneous in structure, sometimes of a slightly pale colour, more frequently colourless, possessing feeble refracting powers, and in many ways suggesting a fatty nature. This, however, is not borne out by chemical tests, which have failed in my hands to demonstrate its actual nature. I can best compare the appearance of these particles to the spores of the favus fungus. It, like the hæmatoidin deposits, has a tendency to aggregate at the bifurcations of vessels, which is no doubt due to the sacculation of the hyaline membrane at these points. This molecular material is greater in quantity and darker in colour, in proportion to the length of time during which disease has existed. It has been found very sparsely scattered over the vessels of sane subjects who had died of fever, and in one case of pneumonia.

When we harden and cut thin sections of the brain of a subject whose insanity had been recent and acute, we find the vessels full of blood, the corpuscles closely packed together, and the coats applied to the retaining cylinder, or, as frequently happens, a vessel is actually kinked or twisted on itself. In chronic cases, on the other hand, and more especially in epileptics and general paralytics, the transversely cut vessel is seen surrounded by a clear ring of unoccupied space, with radiating trabeculæ of connective tissue extending between the hyaline membrane and the cerebral substance. In extreme cases the cylinder has been found from four to six times the calibre of the contained vessel. From experimentation I have found that when the cerebral vessels are thoroughly congested, as after strangulation, their coats are closely applied to the cylinder; and when they are emptied of blood, as after death from hæmorrhage, that a space exists between them. It would appear from this that in health the vessel pretty nearly fills the cylinder, only

[1] *Vide* Bucknill and Tuke's Psychological Medicine, plate ix. fig. 2.

allowing room for the passage of lymph products by the perivascular lymph spaces, and that the clear spaces I speak of are caused by the dilatations of the congested vessel as it produces the expansion of its surroundings. On the resumption of the function of its muscular coats it contracts to its original calibre, leaving a space, which becomes the receptacle of exudates thrown out in subsequent congestions, and so, gradually, the non-contractile brain substance is pushed further and further from the vessels. These spaces have been called perivascular canals, and have been mistaken for the perivascular lymphatic spaces, which you will remember are between the hyaline membrane and the outer fibrous coat.

In the proper coats of the arteries, various degrees of thickening are the most common lesions. As regards the fibrous coats, the inner or intima is much more frequently the seat of this condition than the outer. Hypertrophy of the muscular coat is an almost constant appearance presented in chronic cases, and the circular fibres are, as far as I can ascertain, the ones affected. It is only present in old-standing cases, and is unassociated with hypertrophy of the heart or valvular disease of that organ, or with any kidney affection. As to its causation, we are precluded from any theory founded on a propelling function of the muscular coat, as physiologists are agreed that this is not the function of the muscular fibres of arteries. Although it does not come home to our preconceived ideas that the over-exercise of inhibitory or controlling function is likely to result in hypertrophy of the tissues which exercise it, we have no reason to believe that such is not the case, and the hypertrophy of the muscular coats of the cerebral arteries in the insane affords support to such a theory. In them a constant call is made for an endeavour to maintain normality of blood-supply—that is to say, they are called on in a degree greater than in absolute health to work, and in over-work they become hypertrophied like any other muscles, whatever their function.

It was not my intention to speak of the morbid appearances presented in the brains of subjects who had died from any particular form of insanity, merely to keep to the general question; but within the last few days I have met with such important lesions in the brain of a patient, the subject of what I believe was true syphilitic insanity, that I feel compelled to depart from my proposed plan. In sections made from a subject of this disease, you will notice the arteries more or less thickened as to their muscular coats, and more especially as to their outer fibrous coats—this latter coat is to be seen surrounding the vessels in concentric rings for a very considerable extent, and appears to be cemented as it were by a viscid gummatous material in which are occasionally seen amyloid bodies.[1] The effect of this thickening is completely to occlude the vessels in many instances, and in all very materially to modify their calibre. In the case from which these preparations were made,

[1] *Vide* Journal of Mental Science, October 1874.

the clinical history was perfect; on post-mortem examination, local softenings were found sufficient to account most satisfactorily for a long series of physical symptoms, and the external arteries were most markedly affected by an advanced syphilitic atheroma. The clinical history of this case will be found in the January number of the "Journal of Mental Science," and the full post-mortem observations will be published in the next issue. I will only mention here that the thickening of the arteries was best marked in the neighbourhood of softenings. This obliterative thickening of the arteries of the brain is most interesting when viewed by the light of the observations of Oedmansson and Fränkel on the condition of the arteries of the villi of the syphilitic placenta. Their description and plates indicate an identical condition.[1] These authors hold that in the placenta the thickening appears on the outer layer of the intima, and extends outwards and inwards until the calibre of the vessel is obliterated. Likewise, Byrne and Verdier consider that an obliterative thickening of the branches of the umbilical artery in the placenta is characteristic of syphilis, to which is attributed the frequent death of the foetus.[2] The collation of these observations, so thoroughly independent, and involving such different structures as the brain and the placenta, presents a very beautiful and interesting pathological homology.

When speaking of normal histology, as to the prolongation inwards of the pia mater, as the hyaline wall of the lymphatic spaces which exist around the cerebral vessels, a thickened condition of it was used for purposes of demonstration. In all cases of long-standing insanity, and in old subjects who had not been insane, this membrane, which is in health very fine, perfectly hyaline, and diaphonous, becomes opaque and fibroid, presenting apparently longitudinal fibres.[3] The fibroid appearance is, however, due to puckering. It is in this condition that the hyaline membrane is so easily demonstrable. Last year a discussion raged for months in London, as to the existence of a hyaline fibroid membrane on the vessels of the pia mater of subjects who have died from the cirrhotic or gouty or contracting form of Bright's disease. Sir W. Gull and Dr Sutton have figured such an appearance, and Dr George Johnson pointed out that it can be produced by the application of certain reagents. With all the deference due to such high authorities, I may state my belief that they both are right and both are wrong. I believe that Gull and Sutton are right in asserting that they have seen a thickened membrane outside the outer fibrous coat in cases where the muscular coat was hypertrophied, but that they were wrong in supposing it to be a peculiar characteristic material—a product of a specific morbid

---

[1] *Vide* Archiv. für Gynaekologie, Band v. Heft i.
[2] *Vide* Recherches sur l'Apoplexie Placentaire de Paris, 1868.
[3] Bucknill and Tuke's Psychological Medicine, plate ix. fig. 3.

process, arterio-capillary fibrosis. Nor can I admit that Dr G. Johnson is correct in asserting that this so-called hyaline fibroid membrane is due to the action of chemical reagents, such as glycerine, acetic acid, camphor water. It is true that these agents do bring it more perfectly into view, by occupying the space between the vessel and the membrane, while they at the same time clear up the general opacity of the superficial parts; but the hyaline membrane can be easily made out by the simple use of water alone, and in prepared specimens can be found separate and distinct from the other vascular structures.

The cerebral vessels are frequently found twisted or kinked, somewhat like what happens in overstrained cords; and the cause is probably the same in both, i. e., the tendency which overstretched fibre has to kink when relieved from tension.

I have spoken of the morbid conditions of the vessels nearly in the order of their incidence. 1st, Simple dilatation; 2d, Exudate deposits; 3d, Opacity and thickening of their hyaline membrane; 4th, Dilatation of the retaining canal; and, 5th, Hypertrophy of the muscular coats. This is of course the sequence which might have been expected, but it has been proved by the observation of a very large number of specimens, taken from subjects whose full clinical history had been ascertained.

In rare instances, aneurismal dilatations have been found, but only in the corpora striata.[1]

Degeneration of Nerve Cells.

Considerable interest centres around the study of degenerations in the nerve cells, in consequence of the subtle and peculiar nature of their conformation. The study is fraught with very considerable difficulty, in consequence of the delicate nature of their structure, and the fineness of their processes and poles; and when this has been overcome a still greater difficulty presents itself to the pathologist, namely, that of determining whether these degenerations are of primary or of secondary incidence—that is to say, whether the degeneration which is seen upon the cell is an indication of a primary degeneration of the cell, potential in the perversion of blood-supply, or whether the appearances presented are merely secondary and sequential.

In the present state of our knowledge, this question must be left unanswered, as we have too few data whereon to base a definite opinion. Arguing from the analogies of the state of the cells of the ganglia of nerves supplying regions, the seats of localized paralysis, and the atrophy of certain cells in the spinal cord consequent on amputation of the extremities, the probability seems to be, that the lesions partake of both characters, that there are efficient and resultant abnormalities of the cells of the hemispheri-

[1] Bucknill and Tuke's Psychological Medicine, plate ix. fig. 4.

II

cal ganglia. I will now briefly mention the departures from
normality which have as yet been detected, and in the concluding
lecture indicate the few facts which seem to have a bearing on the
point at issue.

When we examine a recent specimen, taken from the convolutions of an old-standing case of insanity, it is not uncommon
to find the cells represented by a fatty or oily-looking material, retaining in some degree their original shape, but which can be easily
broken up by slight pressure into a confused debris ; this appears to
be the extreme morbid condition, for in the same specimens, cells can
be found covered with a mass of fine highly refracting material,
destroying the definition of the angles, and blurring as it were
their contour.

The nature of this material it is difficult to determine, for the
most careful application of oil tests has failed to produce any
change in its condition or general appearance. When similar
specimens have been submitted to hardening agents, and cut into
thin sections, little change in the morbid cell is seen different
from what presents itself in the recent condition.

Standing upon the table, there are several sections which illustrate
very distinctly not only a molecular, but a pigmentary, condition of
the cells. One of these is taken from the parietal lobe of a man who
died three weeks after the incidence of violent melancholic mania, but
who, for many weeks prior to the actual outbreak, had suffered from
psychical influences of the most exhausting character to the nervous
system. A glance will show that the large cells of the fourth and
sixth layers are those which are affected, whilst those of the external layers are, to all appearance, intact. Of this case I should like
to observe that much less evidence of cellular degeneration was
observable in the frontal and occipital lobes than in the parietal.

Other sections before you present cells in an extreme condition of
granular pigmentary, or, as it was called in the first paper written
by Dr Rutherford and myself, *fuscous* degeneration ; in them the
cells have almost completely lost their normal appearance, and are
represented by rusty-coloured masses, nucleus and nucleolus
having entirely disappeared.[1]

These molecular deposits generally appear first at the base of the
cell, although they are occasionally to be seen surrounding the
nucleus in the first instance, but under either circumstance this
body (the nucleus) is the last to be affected; indeed, it is not
very rare to see a nucleus denuded of protoplasmic surrounding,
and lying naked in the neuroglia.

In the subjects of senile insanity, atrophy of the cells of the outer
layers of the convolution is often a well-marked condition ; they are
shrunk to a fourth of their natural size, are surrounded by a clear
space, larger than exists in health, and their nucleolus shows as a black
spot. In similar cases, and in others of old standing, in which re-

· Bucknill and Tuke's Psychological Medicine, plate x. figs. 4 and 5.

mittent or chronic mania had been the prevailing symptom, and in which the pia mater is so closely adherent as to render it impossible to strip it off without tearing the subjacent cerebral tissue along with it, an all but total obliteration of the cellular elements has been noted; along with this is associated a general coarseness of all the constituents, the definition of the layers is destroyed, and the organ in no manner presents its normal histological appearance.

There is, in contradistinction to this atrophy, a condition in which the cells of the fourth and sixth layers are found large and dilated, puffed out as it were, so as to render them rounder and larger than in health.[1] The nucleus and nucleolus are well pronounced, and no evidence of disease further than the enlargement can be detected.

Apart from the actual condition of the individual cell, we have departures from its normal relation. You will remember that the specimens of normal brain tissue submitted to you in the first lecture, showed a most beautiful and regular fan or fountain-like arrangement. In old-standing cases of insanity, this demonstration can never be obtained; the cells are distorted as to their direction and general arrangement. The cause of this will be seen when we come to speak of the lesions of the neuroglia.

LECTURE VI.

In the last lecture, it was said that the great majority of brain disturbances are due to changes in the blood-supply; but there exists, undoubtedly, a powerful minority in which we have no right to presuppose any such condition, in which we find no evidence of stasis. In certain of the pathological forms of idiocy, which have been so ably described by Dr Ireland of Larbert, various factors have been at work to establish defects, which in the individual are so far normal; that is to say, his brain is so constituted congenitally or by the accidents of childhood, that in him the cells and fibres are so mal-arranged, their constituent elements so imperfect, as to prevent the possibility of the exercise of psychical function to the extent of what is acknowledged to be requisite for a responsible member of society. But this condition in the individual is not conditioned by any peculiarity in his vascular arrangements—he is simply an imperfect human being; he is only relatively abnormal; his defective brain is but part and parcel of his otherwise enfeebled physique,—on a par with his half-withered limbs. The idiot is the extreme case, but there are idiots and idiots. What can the pathologist say about the idiot who manifests his idiocy simply by perversion of the moral sense,—the physically healthy man or woman who steals, secretes, lies, with no disturbance of his intellectual faculties, except that he cannot appreciate that he does wrong in lying, secreting, and stealing—he cannot, as it were, disapprove of

[1] *Loc. cit.*, plate x. fig. 4.

his own proceedings? I am free to confess that no explanation of this section of insanity can be as yet afforded by morbid anatomy. Argument from analogy only warrants us in presupposing a deficiency or mal-arrangement of one or other of the brain elements. This condition must in the meantime be relegated to the somewhat long category of *opprobria medicinæ*. I am not desirous of begging the question; ignorance is simply admitted as to the physical cause of a morbid condition.

Taking a further step, can we give a pathological reason for those rare cases of simple melancholy without excitement which occasionally present themselves? or of those in which a simple condition of delusion is the only symptom? I would only ask you to remember how very seldom such cases are met with without some evidence of brain and general constitutional disturbance: for my own part, I am somewhat sceptical as to their separate existence.

But there is yet another section of brain lesions which may be produced without any causating modification of blood-supply. I allude to changes in the amount and nature of the neuroglia, changes which are due to morbid processes commencing primarily in its own constituents: these may be, in fact generally are, followed by vascular disturbance, and accordingly great difficulty exists in determining whether they are causes or effects. We have to depend mainly on the clinical history of the case in forming an opinion. I do not propose now to go into symptomatology in any detail, but will merely say generally, that when the earliest symptoms present a gradually increasing apathy, torpor, and stupidity, quickly followed by general bodily symptoms, loss of expression, heaviness and immobility of the facial muscles, a slobbering mouth, speech thick and slow, impairment of reflex action, considerable general anæsthesia, hæmatoma auris, and the passage of fæces and urine apparently involuntarily, we may look to the neuroglia as the seat of primary lesion. It is a condition of true primary dementia, which may perhaps best be described as in every way lacking acute symptoms, but accompanied by rapid decadence of the general system. It is closely allied to what is often spoken of vaguely as organic brain disease. In all such cases we find a sclerosis, and generally a colloid degeneration of the neuroglia.

But undoubtedly the neuroglia is subject to certain changes due to congestion and stasis, which must be regarded as the cause of chronic insanity. We have a terminative as well as a primary dementia; the former is infinitely more common, but may be referred to a condition or conditions somewhat similar to what occur in the primary form of the connective tissue, in addition to lesions of cells and fibres. But before considering the period of incidence, we will as shortly as possible discuss what are the morbid conditions of the neuroglia.

When we remember the differences of consistence shown by various brains immediately on their removal from the skull as tested merely by the sense of touch, we cannot fail to recognise, even by this imperfect test, that the connective tissue or packing material, as it may be called, must be the element which suffers in such conditions as œdema, hypertrophy, or atrophy. This is confirmed by microscopic examination, for by this means we are able to demonstrate many various forms of degeneration in the neuroglia. One of the processes to which these sections are submitted, that of staining with carmine, produces naked-eye evidence of a healthy or unhealthy condition of this substance; for if a section takes on a very deep colour, it may be taken for granted that there is an increase of the fibrillar element of neuroglia, and if the converse happens, that its protoplasm is changed in either quantity or quality.

The following are the different forms of disease to which the neuroglia is liable:—First, a general increase of its protoplasmic material; second, a disseminated sclerosed condition; third, a form of sclerosis which has been termed miliary sclerosis; fourth, an increase of its finely fibrillated texture; fifth, a colloid form of degeneration; and, sixth, atrophy. A most interesting instance of the first-named condition came under my notice about eighteen months ago, in a congenital idiot,[1] whose brain weighed sixty ounces: there was a manifest difference between the bulk of the two hemispheres, for, when carefully separated, the right half was found to weigh 30¼ ounces; the left, 23½ ounces; a difference of 6¾ ounces. It would be uncalled for now to enter on the full details of this case; suffice it to say, that the left side of the body was affected by atrophic hemiplegia. In the hypertrophied hemisphere (the right), the nerve fibres were seen lying in fasciculi, consisting of four, five, or six strands, separated from one another by very clear plasm, in which were more than normally numerous nuclei, and a firm fibrillar structure. This condition was best marked in the occipital lobe, less so in the parietal, and in a still minor degree in the frontal; in fact, it was coexistent with the degrees of hypertrophy of the several lobes. It may be regarded as a general or diffused glioma, and to its presence may be due the bulging brain of certain epileptics; and I would further suggest the probability that the weighty brain, not unfrequently met with in idiots who in infancy had been the subjects of hydrocephalus, may be mainly composed of this material. It must be admitted, however, that further proof is needed to establish this theory.

The second form of sclerosis is that which has been described by Rokitansky, Leyden, Charcot and Bouchard, Rindfleisch, and others, as gray degeneration, or more lately as *sclerose en plaques disseminées*. It is manifest to the naked eye as light gray tracts of various shapes and sizes, which occasionally present themselves in sections of brains of chronic dements; it is more frequently observed in the motor tract, in the pons,

[1] See Journal of Anatomy and Physiology, vol. vii. May, 1873.

medulla, and in the spinal cord, than in the white matter of the hemispheres; where, however, it does occasionally occur in widespread areas. When it is present great proliferation of the nuclei always exists, and we seldom fail to demonstrate its contiguity to a vessel or vessels around or upon which nuclei are likewise proliferated. These opaque tracts cannot be cleared up by any known agent; in them nerve fibres are to be seen partially or completely atrophied according to the stage of the disease; in transverse sections the axis cylinders are indiscernible, and the field of the microscope is occupied by a finely molecular and fibrillated material embedded in a cloudy homogeneous plasm. In this matrix the proliferated nuclei exist somewhat enlarged, and atrophied nerve fibres occasionally project raggedly into it. It appears to me that the views of Rindfleisch are most probably correct as to the genesis of the lesion : he holds that the nuclei on and around the vessels first become proliferated, that this is followed by increase of the nuclei of the neuroglia, and the development of a morbid plasm, which is in all probability modified connective tissue.

The disease termed miliary sclerosis was first described by Professor Rutherford and myself in the Edinburgh Medical Journal for September 1868. Its presence has since been noted in various forms of nervous disease by Dr Kesteven, Herbert Major, and others; and I observe that it has lately been claimed as an undescribed pathological condition by Dr Hun, Pathologist of the New York State Asylum, which to me is gratifying, as confirmation of my observations by a perfectly unprejudiced and independent observer.

Miliary sclerosis is a disease of the nuclei of the neuroglia, and its progress is marked by three stages. In the first, a nucleus becomes enlarged and throws out a homogeneous plasm of a milky colour, and apparently of a highly viscid consistence, for the long axis of the spot is almost always in the direction of the fibres, which are displaced by its presence instead of being involved in it; thus indicating that its density is considerably greater than that of the cerebral matrix. In the centre of these semi-opaque spots a cell-like body is generally discernible, possessing a nucleus ; this is the original dilated nucleus of neuroglia. In the largest patches more than one cell can be seen ; whether these arise from division of the first nucleus involved, or from the original implication of more than one nucleus, has not been determined; but from the fact that multiple cells are seen only in the largest spots, it is most likely that the latter hypothesis is correct. Occasionally several neighbouring nuclei become diseased simultaneously, coalesce, and form a multilocular patch of considerable extent; the largest spot which has yet been figured is the one-fortieth of an inch in its longest diameter. During the second stage of development the morbid plasm becomes distinctly molecular in character and permeated by fibrils. It is probable that at this period a further displacement of the contiguous tissues takes place, as a degree of induration of the compressed fibres and bloodvessels which curve round the diseased tract is indicated by the increased amount of colouring material which they then absorb. At this stage the morbid tracts present the following appearances. As a rule the spots are unilocular, occasionally bilocular, and in rare instances multilocular; but whatever their condition in this respect is, they possess the same internal characteristics. A thin section prepared in chromic acid viewed by the naked eye shows a number of opaque spots irregularly distributed over the surface of the white matter; they are best seen in a tinted section, as they are not colourable by carmine. When magnified by a low power they have a somewhat luminous pearly lustre, and when magnified 250 and 800 diameters linear they are seen to consist of molecular material, with a stroma of exceedingly delicate colourless fibrils. They possess a well-defined outline, and the neighbouring nerve-fibres and blood-vessels are pushed aside and curve round them. In well-advanced cases the plasm seems denser at the circumference of the spots than at their centre, and a degree of absorption of the contiguous nerve-fibres is evident; this solution of continuity is only noticeable at the point where the lateral expansion is greatest.[1] The spots are generally colourless, but in some instances they are of a yellowish-green tint, which may be attributable to chromic acid. They vary much in size; multilocular patches are 1-50th of an inch to 1-100th of an inch in diameter, the unilocular ones from 1-200th to 1-800th of an inch. As many as eleven locules have been noticed in one patch, separated one from the other by fine trabeculæ of tissue.

The nervous tissue of a section containing spots of miliary sclerosis in the second stage, when removed from spirit and allowed to dry, shrinks from the diseased patches and leaves them elevated and distinctly separated from it, so much so that they can be picked out with the point of a knife.

It is still doubtful whether this lesion can be detected in recent specimens. In two cases, in which, by the chromic acid process, miliary sclerosis in the second stage was demonstrated, the recent white matter, when squeezed out under a covering-glass, exhibited spaces containing a clear material in which some rounded nucleated cells were visible. It is, however, only by prepared sections that its presence can be definitely ascertained.

In the third stage of miliary sclerosis the molecular matter becomes more opaque and contracts on itself, the boundaries become puckered and irregular in outline, and the material often falls out of the section, leaving ragged holes. These holes cannot be mistaken for empty perivascular canals, which, whatever their size, are smoothly rounded or oval. When in this condition the morbid products of miliary sclerosis are distinctly gritty, effervesce immediately on being subjected to the action of nitric acid, which

[1] See Bucknill and Tuke's Psychological Medicine, plate x. figs. 2 and 3.

produces no such appearances as are evolved by its application in the second stage.

The very fine fibrillar stroma of the neuroglia, which in health can with difficulty be distinguished by the aid of the most powerful lenses, often presents itself in thin sections prepared from the brains of the insane as a distinct meshwork under comparatively low powers. The spider-like cells of Deiters show themselves with great clearness. This condition is best marked in general paresis, and more especially at and near the line of demarcation between the white and gray matters. I see that Subimoff claims this as a new discovery, but it has been well known in this country for some years. This reticulated state of the neuroglia appears to have a great tendency to displace the cells and fibres, which become aggregated, with spaces intervening, composed of altered connective tissue. It must be regarded as a very important and interesting brain lesion.

Lastly, we come to a condition of the nuclei of the neuroglia to which I have been in the habit of applying the term colloid degeneration. I do not think it is a good term, but have sought vainly for a better. It is one of the most important degenerations of the neuroglia, not only on account of its degree, but of the frequency of its presence. It and miliary sclerosis are the lesions spoken of as ever present in true primary dementia

In its earliest stages this abnormal condition shows itself in circumscribed semi-translucent spots scattered irregularly over the surface of the section, varying in size from the 1-4000th to the 1-2000th of an inch in diameter; they have well-defined irregular edges, and their contents are molecular in appearance. In fresh specimens, however, this molecular appearance is not observable, and colloid bodies appear as round or oval in form, having a distinct wall containing a clear homogeneous, transparent, colourless plasm, and occasionally showing a small nucleus, but no nucleolus. Colloid bodies are not colourable by carmine. They appear first in the white matter immediately contiguous to the cortical substance, but as the disease advances they become diffused outwards and inwards. In extreme cases, the appearance of sections containing them may best be compared to a slice of sago-pudding,[1] for they exist in such large numbers as almost completely to fill the field of the microscope, separated slightly from each other by a fine granular material. Although readily recognisable when set up in Canada balsam or turpentine, the characteristics of colloid degeneration are best brought out by glycerine. I feel strongly inclined to regard this as a form of degeneration of the nuclei of the neuroglia; it is first seen and is best marked in the white matter, but in certain specimens in which it occurs in the gray matter cells have been seen which are undergoing, or have undergone, changes in many respects resembling those noticed in the nuclei of neuroglia. It is

[1] See Bucknill and Tuke's Psychological Medicine, plate x. fig. 8.

not associated with proliferation of nuclei. Careful study of a large number of specimens leads to the conclusion that the nuclei are the original seats of the disease, for in all cases in which colloid degeneration shows itself they are to be seen more or less departing from normality; in fact, it may be safely stated that they are always unhealthy, and appear to merge gradually into the colloid condition.

There is one morbid product found in the brains of the chronic insane, and in the spinal cord of epileptics, of which I can offer no explanation as to its origin. These are the amyloid bodies of Virchow. He appears to believe that they are normal constituents; but when you compare the various preparations on the table, it is difficult to understand how the position can be maintained, more especially as they have never been found in any section of healthy nerve tissue which has come under my observation, or that of any histologist who has written on the subject. Like colloid, amyloid is not a good term. They are to be met with in the substance of the convolutions and cord, but their chief seat appears to be the extreme periphery, bound down by the pia mater.[1] I met lately with a very beautiful instance of this condition in the case of a man who, for forty years, had been the subject of chronic mania. The outermost layer of the gray matter was represented by a distinct stratum of these amyloid bodies, which were diffused very generally over the whole surface of the convolutions retained *in situ* by the pia mater. It can only be suggested that they are a modification of the granules of the granular layer.

I must refrain from speaking of many forms of lesion, which, although interesting, do not appear to have such direct bearing on psychical health as those just described. We will therefore, in a few sentences, consider the probable influence which the various lesions described have on the arrangements of the constituents of the hemispherical ganglia. Enough has been said at various times as to the primary effects produced by modifications of blood-supply in the induction of acute insanity, and I will now only refer to degeneration of the cells and of the neuroglia resulting from it, as influencing the chronicity of insanity. In the first place, when we consider that there is strong evidence before us that the vessels of the brain are for long periods dilated far beyond their normal calibre, and when we remember how very highly vascular an organ it is, it is self-evident that there must be loss of cerebral tissue concomitant with the increase of blood. This is to some extent provided against by the displacement of the cerebro-spinal fluid, but this fluid is far too small in amount to represent a bulk of matter equivalent to, say, a doubled amount of blood in the encephalon; and we are therefore compelled to the conclusion that one or other, or all of the solid structures, suffer in a greater or less degree. The unoccupied space represented in this diagram, as affecting an individual vessel, is the result of a loss of brain substance; and if such a condition

[1] See Bucknill and Tuke's Psychological Medicine, plate ix. fig. 6.

I

extends to all the arteries of the superior convolutions, the total amount of tissue absorbed must be very considerable. We find dilated perivascular canals in both white and gray matter, but, as a rule, much more frequently in the former than the latter—in consequence of the larger size of the vessels which permeate it. That the effect of such a condition of the cortical substance must be very prejudicial to psychical health is certain, inasmuch as its cells, nuclei, and fibres are displaced, and the continuity destroyed. In the white matter the normal relations of the fibres is materially interfered with, and absorption of neuroglia produced. The evidence of this is to be found in sections of a convolution of a confirmed case of congestive insanity, for in such we never find the regular fau-like arrangement of cells and fibres which is the natural condition of health; there are clear interspaces, the cells and nuclei are irregularly distributed, and the fibres are pushed together into fasciculi. Superadded we have the pent-up exudates and products of waste acting in the manner spoken of in the last lecture.

In regard to the microscopic tumours, miliary sclerosis, and colloid bodies, the same remarks apply—the continuity of fibre is destroyed, and foreign bodies are present, implicating the normal anatomical condition.

Gentlemen, were I to enter on the full details of this section of my subject, I should far exceed my limits. The demonstration of the various lesions must suggest to you their probable results. When I commenced to write these lectures, I thought we could easily overcome the subject, but it has widened and widened, and I am compelled to stay my hand at this point. It must be confessed that I have failed in many respects, but I ask you to remember that I have been cramped by the necessity of concentrating my remarks: certain subjects have been slurred over which should have been thoroughly discussed; others have been enlarged on, in consequence of personal interest in them, which might have been deferred. We have but fluttered around the outskirts of this magnificent field of inquiry. But, if I have given you any stronger grounds of belief in the somatic nature of what we call insanity than you previously held, if I have pushed my specialty one degree further into the realm of positive medicine, the main object of these lectures has been obtained.

And now, Mr President and Gentlemen, in bringing my lectures to a close for this year, I must, in the first place, beg your pardon for making promises which the time at my disposal has not enabled me to fulfil. I proposed, in the first lecture, to correlate Ferrier's experiments with the results obtained from pathological examination of the brain of the insane; in my last lecture I promised to take up such an important pathological question as the action of the increase or decrease of the cerebro-spinal fluid on pathological conditions; but the *edax rerum* forbids anything but a superficial review of these questions; and as they are questions of too great dignity to be so reviewed, we must relegate them to the future—and possibly the delay of a year will not militate against the value of the inquiry. We have yet to obtain the full details of Ferrier's further experiments; and the subject of the influence of the cerebro-spinal fluid under morbid conditions could not be undertaken without careful consideration of Leyden's monograph on the subject.

I only ask you now to review with me briefly the general outcome of my lectures. We started with the proposition that insanity consists in morbid conditions of the brain, the results of defective formation or altered nutrition of its substance, induced by local or general morbid processes, and characterized especially by non-development, obliteration, impairment, or perversion of one or more of its psychical functions. Holding this as a proposition to be proved, I sought by every means at my disposal to demonstrate, as a necessary first step, that the arrangement of the constituents of the brain was definite and distinct; that their normal relationships were fairly established; and that their individual histology was as far within our ken as that of the majority of other tissues of the body. As a corollary, it was stated that, in the brain, as in the other organs of the body, the normal exercise of function is dependent on a perfect maintenance of the anatomical relations of the component structures. We next proceeded to consider what were the somatic conditions which involved the brain primarily or secondarily, tending towards a solution of the continuity of its constituent elements. In the last two lectures we have devoted our attention to the abnormalities of structure which are constantly found in the brains of the insane. In this procedure, I claim to have followed the course of the teacher of medicine in regard to other organs of the body; and I further claim to have never, in one single instance, failed to support statements by absolute demonstration, whether in regard to the appearances in health or in disease. This principle has been so strictly adhered to, that many points which appeared almost warrantable to advance as definite statements, have been withheld, because they were only *almost* warrantable.

Have we then asserted for insanity a position to which is due the term, *pathological entity?* In what have we failed? Is it that we have not succeeded in bridging the chasm between pathology and metaphysics? This we have certainly not effected, nor is it demanded of us as physicians. *Ne sutor ultra crepidam* is a good old saying, and is strongly pertinent to our position; for, were the physician to hold his hand till he had thoroughly correlated metaphysics with physics, he could never feel justified in administering tonics to the dyspeptic, or stimulants to the depressed patient. To quote the nervous words of Griesinger, " It has been supposed that the study of mental disease was distinguished by some difficulty *sui generis*, and that the study of ordinary medicine had no direct

72

bearing upon it—that the only entrance to psychiatric lay through the dark portals of metaphysics. And yet the other cerebral and nervous disorders which, with the so-called mental diseases, form one inseparable whole, have not, so far as I am aware, been hitherto in any degree elucidated by metaphysics." It is in the mortuary and the work-room that the arcana of cerebral pathology will be disclosed; the section-knife and the microscope will, at no very distant period, lay open secrets which the iteration of theory by the abstract philosopher can never discover.

Feeling certain that there are points connected with the normal anatomy of the brain which demand elucidation, and which have important bearings on its pathology, I propose to conduct a series of experimental inquiries during the next few months, which I hope may form the basis of next year's Morisonian lectures. Our knowledge of the angeiology of the encephalon is still imperfect in respect of the differentiation between arteries and veins in the cerebral convolutions. The question of an encephalic lymphatic system is an open field of inquiry, and I even experience doubts as to the correctness of the generally recognised views of the relations of the cerebral investments. To the study of these subjects my attention will be directed before I next address you; and when I add that the sympathies of Dr M'Kendrick and Mr James Dewar have been enlisted in a work which demands the co-operation of the physiologist and the physicist, it may be fairly hoped that by next session something more worthy of your attention will have been evolved than what my unaided efforts have as yet produced.

EDINBURGH: PRINTED BY OLIVER AND BOYD.